AF589435

John Reuben

Resistive RAM and Peripheral Circuitry

An Integrated Circuit Perspective

Bibliographic Information:

Book title:
Resistive RAM and Peripheral Circuitry
An Integrated Circuit Perspective

Author:
John Reuben, Ph.D
Chair of Computer Architecture
Friedrich-Alexander-Universität Erlangen-Nürnberg (FAU)
Erlangen, Germany
ORCID: 0000-0002-7891-4975
Scopus Author ID: 55270477700

ISBN: 978-3-00-077848-3

Edition Number:
First Edition, Published February 2024

Number of Pages: 148
54 Color Illustrations

©2024 John Reuben

All Rights Reserved. No part of this publication may be reproduced, stored in a retrieval system, or transmitted, in any form or by any means, electronic, mechanical, photocopying, recording or otherwise, except as permitted by law. The rights of translation, reprinting, reuse of illustrations, recitation, broadcasting, reproduction on microfilms or in any other physical medium, now known or hereafter developed are held by the author.

NO AI TRAINING: Without in any way limiting the author's exclusive rights under copyright, any use of this publication to "train" generative artificial intelligence (AI) technologies to generate text is expressly prohibited. The author reserves all rights to license uses of this work for generative AI training and development of machine learning language models.

Bibliografische Information
(in German):

Bibliografische Information der Deutschen Nationalbibliothek: Die Deutsche Nationalbibliothek verzeichnet diese Publikation in der Deutschen Nationalbibliografie; detaillierte bibliografische Daten sind im Internet über dnb.dnb.de abrufbar.

Die automatisierte Analyse des Werkes, um daraus Informationen insbesondere über Muster, Trends und Korrelationen gemäß §44b UrhG ("Text und Data Minin") zu gewinnen, ist untersagt.

ISBN: 978-3-00-077848-3

Preface

This book grew out of my seven years of post-doctoral research (2017-2023). In the year 2017, I held the Viterbi fellowship in Technion, Haifa, Israel. It was in Technion, I was first introduced to memristors. In those days, ReRAM (a type of memristor) was not manufactured and ReRAM was more a prototypical device, researched in a few labs in the world. From 2018 onwards, I was hosted at the Chair of Computer Architecture, Friedrich-Alexander-Universität Erlangen-Nürnberg (FAU), Germany. At FAU, Germany, I involved in various DFG-funded (German Research Foundation) projects focusing on ReRAM and FTJ (Ferroelectric tunnel Junctions). For ReRAM technology, we used the ReRAM manufactured at Institute for High Performance Microelectronics, Frankfurt Oder, Germany. Hence the book grew out of practical research conducted at some of the leading research labs and universities of the world.

This book is written as an introductory textbook* on ReRAM. Although other books on emerging Non-volatile memories (NVM) are available, there was a need for a specific book on ReRAM, given the market it has. Many existing books (to two of which I myself have contributed chapters) are a compilation of chapters from different research groups. Although comprehensive in coverage, such books lacked a cohesive flow. Moreover, they were written at a higher level since they were summarizing research in this field and were not written as a textbook for a university student. On the contrary, this book is written in an easy-to-understand tutorial manner. The purpose is to clarify basic concepts and introduce the reader to this field of emerging NVMs and ReRAM technology in particular. Hence, this book is written for university students considering a career in semiconductor industry. Although written with university students in mind, practicing engineers and researchers entering this field will find this book

---

* The contents of this book were prepared as course material for "Resistive RAM and In-Memory Computing" course offered at Chair of Computer Architecture, FAU, Erlangen.

immensely helpful because the book condenses the findings of decade-long research in this field into few chapters ($\approx$ 150 pages). A preliminary search into any research database will show that this area of emerging NVM is very dynamic, with approximately 1000 papers published every year. As of 2020, the number of memristor-related articles published in technical literature is around 11000*. Referring to the most reliable technical literature (IEEE Transactions and other top journals), I have not only presented the recent research in a concise manner, but also simplified it for easy reading.

Almost all major semiconductor companies are investing heavily in non-volatile memory design and research. Furthermore, due to the von Neumann bottleneck (the data movement between processor and memory which is costly in terms of energy and latency), there is an increasing trend to move computing to the residence of data *i.e.* memory. In other words, the conventional method of storing data in a memory, fetching them and processing it in a dedicated processor before being written back to memory is being re-engineered. New computing paradigms are being explored which compute or process data at their location, if possible in the memory array itself and its peripherals (in–memory computing) or in a dedicated location near the memory (near–memory computing). In fact, **the computing architectures of future will not distinguish between memory unit and processing unit since they will be closely located or fused deliberately to avoid data movement**. No hardware engineer can afford to be ignorant of NVM since the applications of these memory devices are not just for storage, but also for various types of computing. Therefore, such a book introducing students and practising engineers to the fundamentals of ReRAMs is timely and needed. Finally, ReRAMs and other emerging memories are taught as part of a graduate-level course in semiconductor memories. No prior knowledge of conventional memories (SRAM, DRAM, Flash) is required to understand this book. However, a basic knowledge of electronic circuit design (Analog and Digital Integrated Circuit Design) is needed to understand the peripheral circuitry of ReRAM.

Erlangen, *John Reuben*
October 2023 *Friedrich-Alexander-Universität Erlangen-Nürnberg (FAU)*

* Vlasov, A. I.; Gudoshnikov, I. V.; Zhalnin, V. P.; Kadyr, A. T.; Shakhnov, V. A. 2020. Market for memristors and data mining memory structures for promising smart systems, Entrepreneurship and Sustainability Issues 8(2): 98-115. https://doi.org/10.9770/jesi.2020.8.2(6)

Acknowledgements

My journey with Resistive RAM (a type of memristor) started in the year 2017 in Israel. So, I start by thanking Prof. Shahar Kvatinsky, Technion- Israel Institute of Technology for his wonderful support. I was hosted at his $ASIC^2$ lab in The Andrew and Erna Viterbi Faculty of Electrical and Computer Engineering, Technion, Israel. Special thanks to Eric Herbelin, $ASIC^2$ lab manager for all his help in my time in Israel. Next, I moved to Friedrich-Alexander-Universität Erlangen-Nürnberg (FAU), Germany. I was hosted by Prof. Dietmar Fey at the Chair of Computer Architecture. My heartfelt thanks to Prof. Fey for hosting me and also supporting me in all my research endeavors. I also thank the other members of Chair of Computer Architecture, FAU for their support. Special thanks to Edwin Aures, Lab Manager and the secretarial staff for all their help. Prof. Fey not only provided me with a good environment to conduct research but also nurtured me as a researcher. I sincerely thank him for taking time to review the entire manuscript of this book and giving constructive feedback to improve the manuscript.

I also thank Prof. Christian Wenger, IHP, Frankfurt Oder for collaborating on the research. Prof.Wenger shared his ReRAM expertise with us based on the Resistive RAM devices manufactured at IHP, Frankfurt Oder, Germany. Next, I thank the members of Memristec community in Germany for the various networking events and the collaborative works.

Leaving professional circle, it is time to be grateful to people in my personal life. I thank my family (wife and children) for supporting my research interests and moving with me to different countries and making the necessary adjustments. I thank my parents who decided to have a third child (that's me!) when the Indian norm was two children. I thank my brothers and sisters who added fun to my childhood, each one as different as my five fingers. I thank the God of the bible who sent his son Jesus Christ to be the savior of this world. This life, good health, knowledge and intelligence are gifts from God and I could not have written this book without His unseen hand.

Contents

Part I DEVICE, MODELING AND ARRAY STRUCTURE

1 **Introduction** ... 3
  1.1 Do We Need New NVMs Beyond Flash Memory? ... 3
    1.1.1 Non-Volatile Memory(NVM) ... 3
    1.1.2 Flash Memory and its Challenges ... 3
  1.2 Resistance-switching Memories ... 4
    1.2.1 Resistive RAM: The Focus of This Book ... 6
  1.3 General Scope of the Book: Intended Audience ... 6
  1.4 Circuit Emphasis ... 7
  References ... 8

2 **Resistive RAM Device Characteristics** ... 11
  2.1 Resistive RAM and Memristor: A Clarification ... 11
  2.2 Device Structure and Composition ... 12
  2.3 Switching Characteristics ... 13
    2.3.1 Switching Kinetics ... 15
    2.3.2 Switching Energy ... 16
  2.4 Endurance ... 17
    2.4.1 How is it Measured and Reported? ... 17
    2.4.2 Why is the Endurance Limited in ReRAM? ... 18
  2.5 Retention ... 19
    2.5.1 How is it Measured and Reported? ... 19
  2.6 Variability ... 21
    2.6.1 How is variability Measured and Reported? ... 22
    2.6.2 How is Variability Controlled? ... 22

References ... 23

3 **ReRAM Modeling** ... 27
3.1 What is a Model for a Memory Device? ... 27
3.2 Approaches to ReRAM Device Modeling ... 28
3.3 Stanford-PKU ReRAM Model ... 30
3.3.1 Fitting Algorithm ... 31
3.4 Steps to Simulate a ReRAM Device using Stanford-PKU Model in Cadence Virtuoso ... 36
References ... 37

4 **Array Structure and Memory Density** ... 41
4.1 Memory Array ... 41
4.2 Resistive RAM Array Configurations ... 42
4.2.1 1R Configuration ... 43
4.2.2 1S–1R Configuration ... 44
4.2.3 1T–1R Configuration ... 46
4.3 Selectors for 1S–1R Array ... 47
4.3.1 Required Characteristics ... 47
4.3.2 Types of Selectors ... 48
4.4 Three Ways to Increase Memory Density ... 49
4.4.1 Device Scaling and Scalability ... 50
4.4.2 Multi-Level Cell (Multi-bit) ... 51
4.4.3 3D Stacking ... 52
4.4.3.1 Horizontally Stacked 3D ReRAM ... 53
4.4.3.2 Vertically Stacked 3D ReRAM ... 54
References ... 54

Part II PERIPHERAL CIRCUITRY

5 **Circuits to Read from the ReRAM Array: Sense Amplifiers** ... 61
5.1 Introduction ... 61
5.2 Read-out Circuits: Basic Principle of Sensing ... 61
5.3 Voltage-mode Sensing ... 63
5.4 Current-mode Sensing ... 66
5.5 Time-based Sensing ... 69
5.6 ReRAM technology-specific issues to be considered while sensing ... 71
5.6.1 Read-disturb ... 71
5.6.2 Random Telegraph Noise (RTN) ... 73

5.7 Sense Amplifier Design Example 73
References 75

6 Circuits to Program the ReRAM Array: WRITE Circuits 77
6.1 Introduction 77
6.2 Basic WRITE Circuit 78
6.2.1 Voltage Regulation during WRITE operation 80
6.3 Incremental Program and Verify Approach 82
6.4 WRITE Termination Approach 83
6.4.1 WRITE Termination by monitoring BL current 84
6.4.2 WRITE Termination by monitoring SL current 85
6.5 Writing into 1S-1R arrays 87
6.6 Issues faced while writing into ReRAM cell 89
6.6.1 Write disturb 89
6.6.2 IR drop 90
References 90

7 Row/Column Access Circuits and Multilevel Cell 93
7.1 Row Decoder: Circuit to Select a Row 93
7.1.1 WL driver 96
7.2 Column Multiplexers 96
7.2.1 Predecoding for large arrays 98
7.3 Multi-Level Cell(MLC): Some Directions for Writing and Reading 99
7.3.1 Writing Multiple States 99
7.3.2 Reading Multiple States 101
7.3.2.1 Sense Amplifier for multi-bit data 104
7.4 Memory Architecture: A Brief Introduction 105
7.4.1 Sub-array Design 106
References 108

Part III IN–MEMORY COMPUTING

8 In-Memory Computing: Applications of Resistive RAM Beyond Memory 113
8.1 Introduction 113
8.1.1 Von Neumann Bottleneck (The Memory Wall) 113
8.2 In-memory Arithmetic 116
8.2.1 Boolean Logic Gates in ReRAM array 116
8.2.1.1 NOR Logic Operation in Memory Array 118
8.2.1.2 Majority Logic Operation in Memory Array 119

8.2.2 In-memory Adders 123
8.2.2.1 One-bit Full Adder using In-memory NOR Gate 123
8.2.2.2 One-bit Full Adder using In-memory Majority Gate 124
8.2.2.3 Optimizing Latency of In-memory Adders: Some Directions 126
8.2.3 Eight-bit Parallel-prefix Adder in ReRAM array 129
8.2.3.1 Mapping Methodology 130
8.3 Matrix Vector Multiplication(MVM) in ReRAM Array 133
8.3.1 MVM by Bit Slicing 139
8.4 Conclusion and Future Outlook 143
References 144

Part I
DEVICE, MODELING AND ARRAY STRUCTURE

Chapter 1
Introduction

1.1 Do We Need New NVMs Beyond Flash Memory?

1.1.1 Non-Volatile Memory(NVM)

A non-volatile memory is a device which can retain data without being powered ON. From an electrical engineering perspective, it is a physical device which can store data without being continuously supplied with energy. It must be noted that writing into the device and reading from it does require energy. Energy may be required to write into the device and read from the device, but no energy is required to retain the stored data in the device. Once written, the data is unaltered and remains unaltered for a long period of time, typically many years. For how many years it can retain is defined by a characteristic of NVM called 'retention' and this will be discussed more in detail later.

1.1.2 Flash Memory and its Challenges

Flash memory is a type of non-volatile memory where the information is stored as a charge in the floating gate of a MOSFET structure *i.e.* in addition to a normal gate, the flash memory device has an floating gate to store charge. This presence or absence of charge functions as a memory [1]. Flash memory has been the de-facto standard for NVMs for decades. It has been used in our computers, USB sticks , mobile phones and almost all electronic devices which need to retain some data without being powered ON.

In a textbook on emerging NVM like ReRAM, it is important to first lay the foundation by asking the question - why do we need **new** NVMs in the first place? In other words, is flash memory facing some limitations? Although flash is non-volatile and has been the de-facto standard for large storage, IRDS 2020 report [2] identifies the following inevitable limitations faced by flash memory:

1. Poor endurance* (10^3 to 10^5 erase cycles)
2. Modest retention (typically 10 years on a new device),
3. Long erase time ($\approx$ms)
4. High operating voltage (15 V)

According to Waser et al. [3], flash memory suffers from low endurance, low write speed and high voltages required for WRITE operation. The read, write, and erase latency of flash memory technology have remained essentially constant for more than ten years, despite the technology's projections for greater scaling of density [4]. While the introduction of Multi-Level Cell (MLC) flash devices extended their storage capacities by a small integral factor (2–4), the combination of scaling and MLC have resulted in the degradation of both retention time and endurance, two parameters critical for storage applications [2]. All these challenges has invigorated research in the last two decades for a potential replacement for flash memory.

1.2 Resistance-switching Memories

A resistance-switching memory is any two-terminal resistor device whose resistance can be electrically altered in two stables states [5]. Under voltage/current stress, the resistance can be switched between a Low Resistance State (LRS) and a High Resistance State (HRS) (hence the name resistance-switching). The word 'memristor' is also used by researchers to refer to a resistive-switching device since such a device is basically a 'resistor' with a 'memory'. The word 'memristor' is a broad word and is used to refer to any resistance-switching memory regardless of the underlying device structure

* Endurance denotes the number of times that the device can be switched between two stable states while maintaining enough resistance ratio between them. This will be discussed in detail in Chapter 2

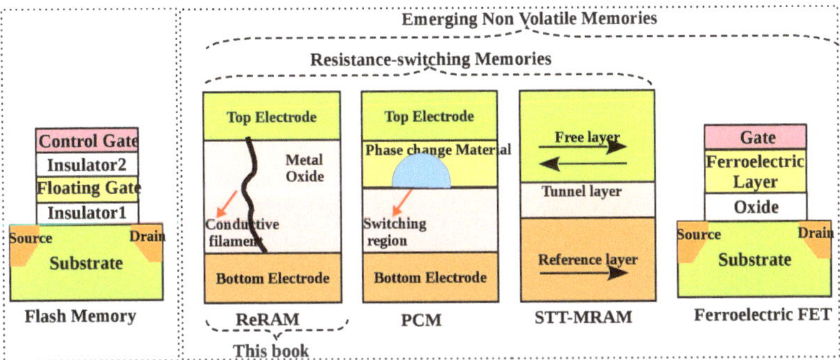

Fig. 1.1 Flash memory (left) and emerging NVMs (right) seeking to replace flash memory, at least partially. ReRAM, PCM and STT-MRAM are resistance-switching memories since the data is stored as resistance. Ferroelectric FET is a type of ferroelectric memory where the data is stored as the threshold voltage of the MOSFET, similar to flash memory (Ferroelectric RAM (FeRAM) and Ferroelectric Tunnel Junction(FTJ) are other types of ferroelectric memories not shown here). The book focuses only on Resistive RAM.

or the mechanism which causes the change in resistance. To differentiate between different types of resistance-switching memories, we need to look at their structure and the accompanying physical mechanism which causes the resistance change. Following this approach, they can be classified as follows:

1. Resistive Random Access Memory (ReRAM) device is a Metal-Insulator-Metal structure where a conductive filament is created (LRS) or broken (HRS) in the insulator. The insulator is usually a transition metal oxide (OxRAM) or an electrolyte (Conductive Bridge RAM)
2. Phase Change Memory (PCM) device is a Metal-Active Material-Metal structure where the active material is a chalcogenide phase-change material which is either in amorphous (HRS) or crystalline state (LRS)
3. Spin Transfer Torque-Magnetic RAM (STT-MRAM) is a Free layer-Tunnel Layer-Reference layer structure where the magnetic polarization of the Reference layer is fixed while that of the free layer can be programmed to be either in the same direction (parallel, LRS) or opposite direction (anti-parallel, HRS)

Resistance switching memories (ReRAM, PCM, STT-MRAM), ferroelectric memories (FTJ, FeFET, FeRAM), macromolecular(polymer) memory, mott memory and DNA memory are some of the emerging memories and each of them are in different stages of maturity. IRDS 2022 report

[6] identifies STT-MRAM, PCM, ReRAM and FeRAM as prototypical memories since they are available as prototypes and even commercially available for niche applications.

1.2.1 Resistive RAM: The Focus of This Book

ReRAM, STT-MRAM and PCM are prominent resistance-switching memories in the early stages of commercialization. All these resistance-switching memories have their own merits and demerits and a comprehensive comparison is not possible since they are in nascency. This book will focus on ReRAM, an emerging NVM having a greater interest and potential compared to other competing NVMs mentioned above. This has been recognised by various researchers and experts in semiconductor memory industry and consequently these memories are in the early stages of manufacturing. Fujitsu manufactures 12 Mbit ReRAM [7]. Infineon is using TSMC's ReRAM in its aurix microcontrollers [8]. Crossbar Inc [9] and Weebit Nano [10] license ReRAM IPs and many other semiconductor memory manufacturers are also entering the market. According to Yole's NVM market intelligence report, emerging NVM market is expected to reach $2.7 billion by 2028, with ReRAM expected to account for more than a third of that [11].

1.3 General Scope of the Book: Intended Audience

Although the book focuses on ReRAM, there is a wider scope for this book in the context of NVM. As a reader or expert in this field of emerging NVM will agree, the basic principles of NVM discussed in the context of ReRAM are applicable for other two-terminal NVM with little or no modifications. For *e.g.*, the chapter on Array Structure discusses different ways in which ReRAM devices are organised into arrays– as single devices (1R) or with a selector(1S-1R) or with a transistor (1T-1R). STT-MRAM is also a two-terminal device which is commonly integrated in a 1T-1R architecture with a magnetic RAM integrated on top of the drain of the transistor instead of the ReRAM. Similarly, the peripheral circuitry (READ and WRITE circuits, row and column decoders) are also very similar and can be adapted to any resistance-switching memory. Hence, this book will be useful for anyone who wants an overview of emerging NVM in addition to those desiring a detailed view of ReRAM technology.

This book caters to various professions. First, it can be used as a textbook for graduate students of electrical/electronics/computer engineering. Most of the universities have a module on memory as part of computer architecture course. The module covers the basics of SRAM, DRAM and flash memory. However, this was the case when flash memory dominated the NVM market for storage class memory. A course on emerging NVM is already beginning to be offered as a separate course in some universities due to the attention these memories have got in recent past. This book can be used as a textbook for an introductory course on ReRAM or used as part of a larger course on emerging NVM. Secondly, practising semiconductor engineers in the field of memory IP development will benefit from the exhaustive coverage of peripheral circuitry. Thirdly, this book will be useful for researchers entering the field of in-memory computing, neuromorphic computing and AI hardware [12, 13], all of which use array of some resistance switching memory as the substrate for processing.

1.4 Circuit Emphasis

Any discussion on electronics can be at different levels of abstraction- device or circuit or system. Any textbook on integrated circuits must make the emphasis clear to the audience. A book on CMOS integrated circuits can place much emphasis at the device level discussing the transistor device physics and the various phenomenon occurring inside that single transistor. Another book can have an emphasis on circuit design - the art of designing circuits using the transistors. A book with system emphasis can elaborate how complex systems like System-on-Chip (SoC) can be designed by stitching together many IP cores. The emphasis of this book is at the abstraction of circuit. In Part I, the author discusses the basic device structure, characteristics, modeling and how these devices are integrated to form a memory array. However, the switching mechanism and the accompanying resistance change are not discussed at the ion level. Neither is device physics the emphasis of the book†. This book views the memory

† The subject of ReRAM device physics and switching mechanism at the device level is well presented in other books. *e.g. Advances in Non-Volatile Memory and Storage Technology*, Woodhead Publishing, 2019, ISBN 9780081025840; *Resistive*

device at the circuit level and gives the minimum device understanding a memory chip designer needs. Given a ReRAM device and its characteristics, how can one design a Random Access Memory(RAM) using it? Part II discusses peripheral circuitry- how to design circuits to read from and write into the memory array? How to design the row and column access control circuits? Part III gives an introduction to the field of in-memory computing. Again, the emphasis is on circuit design- how can ReRAM devices and CMOS circuits be used together to achieve simple computations like addition and matrix vector multiplication? In short, the book tries to answer the questions of a memory circuit designer and not of a memory device engineer.

References

[1] R. Bez, E. Camerlenghi, A. Modelli, and A. Visconti, "Introduction to flash memory," *Proceedings of the IEEE*, vol. 91, no. 4, pp. 489–502, 2003.

[2] International roadmap for devices and systems: Beyond cmos. [Online]. Available: https://irds.ieee.org/editions/2020/beyond-cmos

[3] R. Waser, R. Dittmann, G. Staikov, and K. Szot, "Redox-based resistive switching memories – nanoionic mechanisms, prospects, and challenges," *Advanced Materials*, vol. 21, no. 25-26, pp. 2632–2663, 2009.

[4] L. M. Grupp, A. M. Caulfield, J. Coburn, S. Swanson, E. Yaakobi, P. H. Siegel, and J. K. Wolf, "Characterizing flash memory: Anomalies, observations, and applications," in *Proceedings of the 42nd Annual IEEE/ACM International Symposium on Microarchitecture*, ser. MICRO 42. New York, NY, USA: Association for Computing Machinery, 2009. doi: 10.1145/1669112.1669118. ISBN 9781605587981 p. 24–33. [Online]. Available: https://doi.org/10.1145/1669112.1669118

[5] D. J. Wouters, R. Waser, and M. Wuttig, "Phase-change and redox-based resistive switching memories," *Proceedings of the IEEE*, vol. 103, no. 8, pp. 1274–1288, 2015.

[6] Beyond cmos and emerging materials integration, irds 2022 report. [Online]. Available: https://irds.ieee.org/editions/2022/irds-2022-beyond-cmos-and-emerging-research-materials

[7] (2023) Fujitsu semiconductor memory solution data sheet. [Online]. Available: https://www.fujitsu.com/jp/group/fsm/en/documents/products/reram/lineup/MB85AS12MT-DS501-00071-1v1-E13.pdf (Accessed 2023-10-30).

[8] (2022) Infineon and tsmc to introduce rram technology for automotive aurix tc4x product family. [Online].

Switching: From Fundamentals of Nanoionic Redox Processes to Memristive Device Applications, Wiley-VCH Verlag, 2016, ISBN 9783527334179

Available: https://www.infineon.com/cms/en/about-infineon/press/market-news/2022/INFATV202211-031.html (Accessed 2023-10-30).

[9] (2022) Rethink reram for use as ftp or otp memory. [Online]. Available: https://www.crossbar-inc.com/products/ftp-otp-memory/ (Accessed 2023-10-30).

[10] (2023) Non-volatile memory (nvm) without compromise. [Online]. Available: https://www.weebit-nano.com/technology/overview/ (Accessed 2023-10-30).

[11] (2023) Emerging non-volatile memory 2023. [Online]. Available: https://www.yolegroup.com/product/report/emerging-non-volatile-memory-2023/ (Accessed 2023-10-30).

[12] *Applications of Emerging Memory Technology Beyond Storage*, 1st ed., ser. Springer Series in Advanced Microelectronics, 63. Singapore: Springer Singapore, 2020. ISBN 981-13-8379-0

[13] K. Ishimaru, "Future of non-volatile memory -from storage to computing-," in *2019 IEEE International Electron Devices Meeting (IEDM)*, 2019. doi: 10.1109/IEDM19573.2019.8993609 pp. 1.3.1–1.3.6.

Chapter 2
Resistive RAM Device Characteristics

2.1 Resistive RAM and Memristor: A Clarification

A memristor is a contraction for the combination of memory and resistor *i.e.* a device which is capable of remembering its resistance. The word 'memristor' was coined by Leon Chua in his 1971 paper- *'Memristor-The missing circuit element'* [1]. In this paper, the author stated that such a device is yet to be discovered, but theoretical investigations prove the existence of such a device. Since the word memristor does not say anything about the the device structure or the mechanism which leads to change in resistance, it can be easily misunderstood*. Furthermore, the word memristor is broad and refers to any device which can change its resistance in response to voltage/current stimulation. In fact, all two-terminal resistance switching memories are memristors, regardless of the device material and physical operating mechanisms [2]. Hence, the word memristor is very broad and is agnostic to the underlying device structure or the mechanism which causes the change in resistance.

The word Resistive RAM, on the other hand, is more specific. Firstly, the word ReRAM refers to a two-terminal device in which the data is stored as resistance. Secondly, in a ReRAM device, the change in resistance is due to the motion of electrically active ionic defects (*e.g.* oxygen vacancies or metal cations). This movement of ionic defects results in the formation or rupture of a conductive filament in the device. What makes a resistance switching memory to be classified as ReRAM is the

* In technical literature, the word memristor and Resistive RAM were often used interchangeably and this causes a confusion for a person entering this field.

cause of resistance change *i.e.* movement of ionic defects leading to the formation of a conductive filament in the insulator (HRS $\rightarrow$ LRS) and its rupture (LRS $\rightarrow$ HRS) under appropriate voltage stimulus. Note that STT-MRAM and PCM also exhibit change in resistance under voltage stress, but they are not classified as ReRAM since the change in resistance is not due to the formation of conductive filament. On the contrary, in PCM, the change in resistance is due to change in the material property (from amorphous to crystalline) and in STT-MRAM, the change in resistance is due to polarization alignment between the free layer and the reference layer. Hence, STT-MRAM and PCM can be classified as memristors but not as Resistive RAMs. To summarize, the word memristor is a broad term referring to resistance switching devices and is agnostic to device structure and the physical mechanism which causes the resistance change. The word 'Resistive RAM' or 'ReRAM' specifically refers to devices which store data as resistance **and** the accompanying change in resistance is due to the movement of electrically active ionic defects (which results in formation or rupture of conductive filament). Finally, the suffix 'RAM' in ReRAM is used in the same sense as it is used in SRAM and DRAM *i.e.* it is a memory which permits READ and WRITE operations and the access time for any location in the memory array is the same.

2.2 Device Structure and Composition

Resistive RAM device is a 'Top Electrode–Switching Material–Bottom Electrode' structure, as shown in Fig 2.1. The **state** of the ReRAM device is its resistance *i.e.* the data is stored as resistance and its resistance does not change even after the memory chip is left without any source of electrical power. The resistance can be changed by the application of voltage between the top and bottom electrodes. Under such a voltage stress, a conductive filament is formed or ruptured in the switching material, which leads to change in resistance. Two types of Resistive RAM are distinguished based on the switching material- OxRAM, if the switching material between the two electrodes is a binary metal oxide or Conductive Bridging RAM, if the switching material is oxide or electrolyte (with the top electrode being an electrochemically active metal) [3]. OxRAM and Conductive bridging RAM not only differ in their switching material, but also in the chemical process by which the conductive filament is formed. This is discussed in the next section. Another important characteristic of ReRAM technology is that the switching material and the metal electrodes are not standardized. Different oxides like HfO_2, TaO_2, SiO_2, TiO_2, Ta_2O_5 [4, 5] have been investigated as switching material and different metals like platinum, titanium, aluminium, carbon have been investigated as electrodes [6, 7, 5]. Consequently, ReRAM device characteristics vary significantly between memory manufacturers according to the switching material and its thickness, electrodes used *etc.*

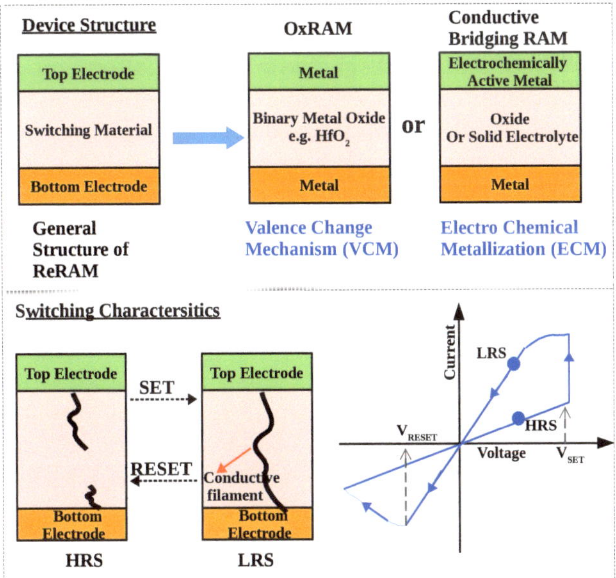

Fig. 2.1 Two terminal structure of ReRAM: OxRAM and Conductive Bridging RAM differ in the composition of the electrodes and switching material and the mechanism by which the filament is formed. In OxRAM, it is by Valence Change Mechanism and in Conductive Bridging RAM it is by Electro Chemical Metallization

2.3 Switching Characteristics

As depicted in Fig. 2.1, the change in resistance in a ReRAM device is due to the formation or rupture of a conducive filament between top and bottom electrodes. The transition from HRS to LRS is called the 'SET' process and the reverse switching event is called 'RESET' process. During the 'SET' process, a voltage $\geq V_{SET}$ is applied between the top and bottom electrodes and a conductive filament is formed between the top and bottom electrodes, thus lowering the resistance (denoted LRS, Low Resistance State). When a voltage of opposite polarity is applied ($\geq |V_{RESET}|$), the conductive filament is ruptured leading to a High Resistance State (HRS). The mechanism of the formation of the conductive filament needs some material science background and is not very relevant here for the circuit designer/ hardware engineer. However, for the sake of completion, the mechanism is briefly stated. In both types of Resistive RAM (OxRAM and Conductive Bridging RAM), the resistance alteration in response to a voltage stress is due to the formation and rupture of a conductive filament. In OxRAM, the conductive filament is formed by redox reaction due to oxygen ion migration [8]. This process is also called Valence Change Mechanism (VCM). In Conductive Bridging RAM, during the SET process, the oxidation of the electrochemically active electrode leads to the formation of cations which drift to the counter electrode under the influence of the electric field, forming a conductive filament [3]. This process is also called Electro Chemical Metallization (ECM). Therefore, OxRAM is also called a VCM device since the resistance switching is due to Valence Change Mechanism. Conductive Bridging RAM is also called ECM device since the resistance switching is due to Electro Chemical Metallization.

The I–V characteristics depicted in Fig. 2.1 is the most common type of switching characteristics – SET and RESET process happen with voltages of opposite polarity. This is called bipolar switching, which is the most common type of ReRAM switching. It must be noted that there also exist ReRAMs with unipolar switching *i.e.* both SET and RESET processes occur with voltage of same polarity. Typically, SET process requires a higher positive voltage and RESET process requires a lower positive voltage across the device. Such unipolar switching is attributed to thermally induced redox reactions activated by Joule heating effects at the conductive filament. However, unipolar switching ReRAMs have lower uniformity (*i.e.* high variability) and lower endurance compared to bipolar switching ReRAMs [9]. Therefore, unipolar switching ReRAMs are not actively pursued and hence not discussed in this book.

Another aspect of ReRAM which may not be very relevant for the circuit designer yet worth mentioning is the concept of 'electroforming'. 'Electroforming' may be seen as the initial step which is needed before it can be 'SET' and 'RESET' and regular switching between LRS and HRS can take place. During this 'initialization' step, the first conductive filament is formed in the pristine device by generation of chains of oxygen vacancies [10]. In other words, the first SET process for pristine devices (freshly fabricated ReRAM samples) is a special step which requires a voltage higher than that required for regular SET process. After this 'electroforming', the ReRAM is ready to be used as a normal memory device. It must be noted that this 'electroforming' is necessary only in OxRAM devices. Conductive Bridging RAM typically do not require this step [11]. Moreover, ReRAM which can function as NVM without such an electroforming step are also being experimented. They are called 'forming-free' ReRAM and the interested reader is referred to [12, 13, 14].

2.3.1 Switching Kinetics

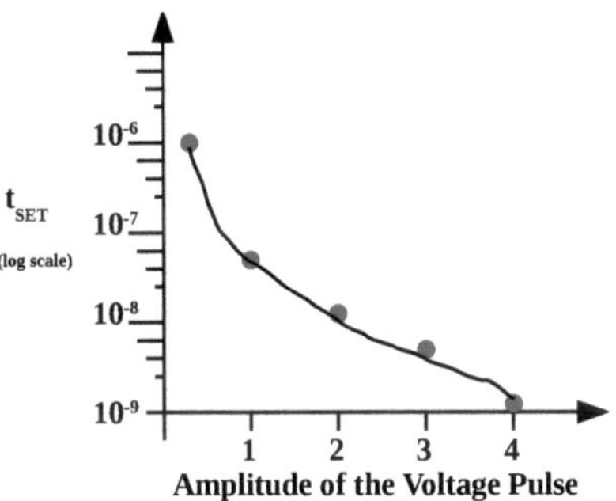

Fig. 2.2 The time to switch a typical ReRAM decreases non-linearly with respect to the amplitude of the voltage applied across its terminals

In the switching characteristic of Fig. 2.1, we have considered only the I–V characteristic of the ReRAM device. When the voltage across the device is varied, how does the current vary? This characteristic does not say anything about how fast the device reacted to this voltage stress. In other words, this simple I–V characteristic does not say anything about the time dimension. However, the switching characteristics with respect to time are also important in certain cases *e.g.* to study the energy spent in switching the device, how fast can we READ and WRITE into the memory array made of such cells *etc.* Researchers have tried to study how fast or slow a ReRAM device can change its state *i.e.* the switching kinetics. If

V_{SET} is the minimum voltage required to switch the ReRAM to LRS, what happens if voltages greater than V_{SET} is applied? Researchers have tried to study this phenomenon by applying pulses of different amplitudes and measuring the switching time. It was observed that the time to switch the ReRAM decreases non-linearly with amplitude of the applied voltage, as depicted in Fig.2.2. It must be noted that Fig.2.2 is a representative plot of typical ReRAM switching time (t_{SET} denotes the time to switch the device from HRS to LRS) with respect to amplitude of pulse applied. This is true of OxRAM devices (observed in Ta_2O_5 [15]) and Conductive bridging RAM devices (observed in SiO_2 [16]). Reader is referred to [15, 16] for exact numbers which differ according to the switching oxide of ReRAM, thickness of the oxide etc. As the amplitude of the pulse applied to switch increases, the switching time decreases exponentially (NOT linearly). This phenomenon is called the non-linear switching kinetics of ReRAM (note that the switching time t_{SET} is in log scale).

2.3.2 Switching Energy

Switching energy of the ReRAM device is the energy required to switch the device from one state to another *i.e.* LRS to HRS or vice versa. In other words, switching energy is the energy spent in programming the device to a particular state (WRITE energy). It is important to have an estimate of the switching energy before investing heavily in a NVM technology because this energy determines the energy consumption of the memory access operations. Since ReRAM is non-volatile, no energy is spent in retaining the state. Energy is spent only when writing and reading from the memory. As we will see in later chapters, the ReRAM device is read by applying ≈ 0.2-0.3 V across it and sensing the read-out current using a CMOS circuit. Hence, the READ energy is less and can be minimized to a large extent. However, during the WRITE process (SET/RESET), there is significant current (tens of μA to 100 μA) flowing through the device and the WRITE energy is more. Consequently, the WRITE energy determines the energy efficiency of the whole memory system. The switching energy during SET process is

$$E_{SET} = V_{CELL} \cdot \int_0^{t_{SET}} I_{CELL}\, dt \tag{2.1}$$

where V_{CELL} is the voltage applied across the ReRAM cell ($V_{CELL} \geq V_{SET}$) and t_{SET} is the duration for which it is applied. Since the device is initially in HRS, the current I_{CELL} through the device is not a constant but varies (increases from I_{HRS} to I_{LRS}) and it is integrated over the SET process duration to determine the energy (a similar process can be used to determine the energy during RESET and the switching energy is the average of the two). The switching energy depends on voltage required to switch (V_{SET}/V_{RESET} which depends on device structure/material), the current during switching (which depends on the HRS/LRS of the device) and the time to switch. Typical switching energy of ReRAM is ≈ 1 pJ [17] and can vary between 0.1 pJ [18] and 50 pJ [19].

2.4 Endurance

Endurance of the memory device denotes the number of times the device can be switched between two stable states. In other words, endurance is a quantitative measure of the stress-handling capability of the device. In the context of ReRAM, by stress we mean the electrical stress *i.e.* the voltage pulse applied to change the state of the device. For ReRAM device, endurance is the number of cycles it can be switched between two stable resistance states while maintaining enough resistance ratio between them. Consequently, if a ReRAM device is said to have an endurance of 1000 cycles, it means that it can be switched 1000 times reversibly between HRS (High Resistance State) and LRS (Low Resistance State). Note that the device switched from LRS to HRS and then back to LRS is considered to be one switching cycle [17], as depicted in Fig.2.3. Consequently, endurance of 1000 cycles means 2000 transitions in total (1000 transition from HRS to LRS and 1000 from LRS to HRS). Endurance is an important characteristic of a memory device since it decides the lifetime of the memory cell.

2.4.1 How is it Measured and Reported?

Typically, endurance is measured by applying a series of SET and RESET pulses with READ pulses between them. Hence, the device is driven from HRS to LRS (SET process), read and then from LRS to HRS (RESET process). During the READ between SET and RESET, the resistance

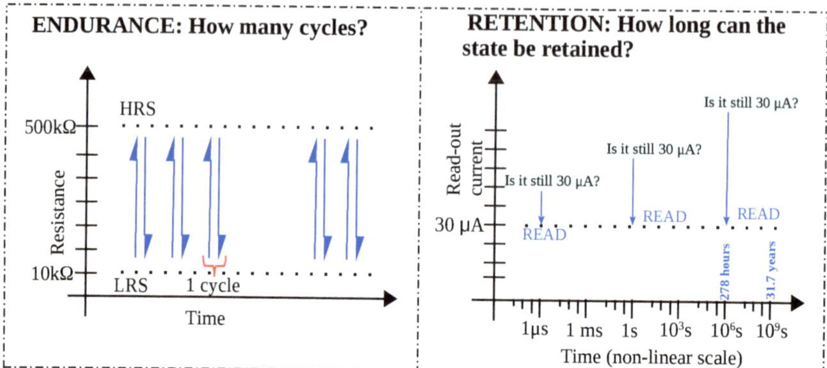

Fig. 2.3 Endurance denotes the number of times the ReRAM device can be switched between two stable states. Retention is a measure of the duration for which the ReRAM device can maintain the state unaltered.

is measured (by measuring the current drawn from the ReRAM at a small READ voltage). This process is repeated for several cycles and the data is stored and analysed to find the time when the discrimination between LRS and HRS fades or gradually disappears. If after n cycles, the memory window is no longer discernible, the ReRAM device is said to have an endurance of n cycles. Hence, measuring endurance is a time-consuming process. In practise, the measurement of endurance is even more complicated due to device-to-device variability *i.e.* such an endurance study must be performed on multiple devices and averaged to obtain a very reliable number for endurance of a ReRAM device.

2.4.2 Why is the Endurance Limited in ReRAM?

The memory window collapses after certain switching cycles, signalling that the device has reached the limit of its stress-handling capability *i.e.* the endurance of the device. In practice, this happens when the device is either stuck at LRS/HRS (failure to switch even when the appropriate pulse is applied) or that the absolute HRS and LRS change significantly over many switching cycles. In both the cases, the memory cell is no more useful since either it does not switch at all or switches but the HRS/LRS deviate significantly from the earlier HRS/LRS. The reasons for this collapse of the memory window are attributed to significant changes in the structure of the conductive filament (also called over-SET and over-RESET in [20]), too many oxygen vacancies [21], oxide hard breakdown [22] and unbalanced SET and RESET pulse amplitudes [23].

As stated, ReRAMs are being fabricated with different switching oxides, oxide thickness and device area. Consequently, they have diverse memory characteristics and many university and industry research groups have reported different endurance for their ReRAM devices. Although endurance from 10^5 to 10^{12} have been reported for various devices [24], the most reliable data from literature suggests that ReRAM endurance varies from 10^5 to 10^9 cycles [25, 26, 19].

2.5 Retention

Retention is the length of time the ReRAM cell can retain the stored data *i.e.* the duration between writing the data and the first erroneous read-out of that written data. Each non-volatile memory has a distinct physical mechanism which is responsible for its non-volatility and the retention characteristic depends heavily on this physical mechanism. In Resistive RAM, the non-volatility is due to the formation and rupture of the conductive filament in the switching oxide. Hence, ReRAM retention depends on stability of the formed filament (responsible for LRS retention) and stability in the filament discontinuity (responsible for HRS retention). In HRS, the filament is broken and it is easier to remain in the state, the natural state of the device. In LRS, the filament is formed and the filament must continue to be stable for a long time (thereby forming a conductive path in the switching oxide), which is a challenge. In LRS, the conductive filament degrades due to unwanted migration of oxygen vacancies or the metal ions which form the filament. Hence, ReRAM retention is dictated more by LRS retention than HRS retention [17].

2.5.1 How is it Measured and Reported?

Retention is normally measured by Constant Voltage Stress (CVS) where a low READ voltage is applied to the device and the state of the device is monitored (by observing the read-out current) to ascertain how long it can retain the state. It has been observed that ReRAM retention is a strong function of the limiting current (also called Compliance Current†) used while switching. Indeed, better retention has been experimentally observed if a larger current was used while switching. This is because, a large switching current forms a thicker conductive filament, which remains more stable over time. Therefore retention data are often plotted along with the switching current, as reported in [17, 5].

The industry requirement for retention is 10 years at 85°C. Since it is not practical to wait for 10 years to verify if a ReRAM cell can retain data 10 years after an initial programming operation, we depend on accelerated tests, which simulate the phenomenon which causes the degradation of the state. In ReRAM technology, typically high temperatures are used during CVS tests to verify retention since at high temperatures, the atoms in the switching oxide acquire energy which facilitates atomic rearrangements [17]. Therefore, the retention measured in LRS at room temperature will always be larger than that measured at higher temperatures. The current estimate for ReRAM retention is 10 years at 85°C [26, 19].

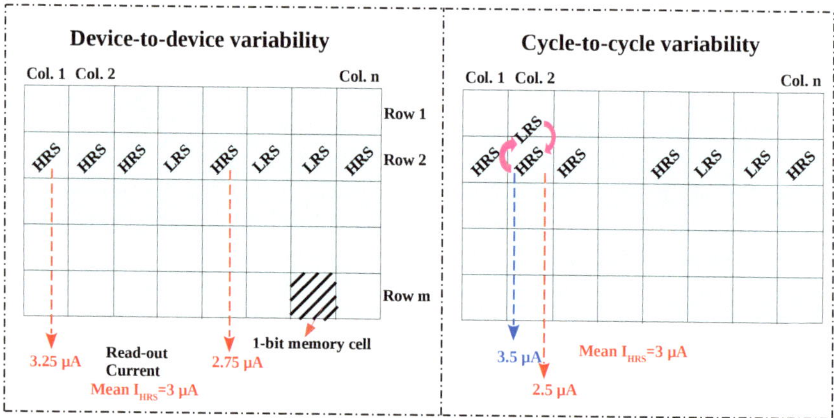

Fig. 2.4 ReRAMs exhibit two types of variability: Device-to-device variability is the non-ideality in the spatial domain, while cycle-to-cycle variability is the non-ideality in the temporal domain.

† CC is implemented by using an access transistor in series with the ReRAM device, which limits the maximum current through the device to avoid hard breakdown while switching from HRS to LRS

2.6 Variability

Variability refers to non-idealities exhibited by a ReRAM device while reading or writing into it. Broadly speaking, variability can be spatial or temporal in nature. By spatial, we mean the variability that is observed between memory devices in the memory array *i.e.* the read-out current of device A in the array may be different from the read-out current of another device B programmed to the same state in the same memory array. This is also called as device-to-device variability (Fig. 2.4). A device programmed to HRS may output a read-out current I. Now, when this device is switched to LRS and then switched back to HRS, it may output a read-out current I'. This is called cycle-to-cycle variability, as illustrated in Fig.2.4.

Variability in ReRAM devices is widely recognised as a serious concern hindering its industry adoption and commercialization [27, 28, 29, 30, 31]. In ReRAM technology, variability in switching behavior can be temporal (cycle-to-cycle) and spatial (device-to-device). The variability in the switching behaviour manifests as variation in the programmed resistance of the device (LRS and HRS) and variation in the voltage at which the device switches (HRS $\rightarrow$ LRS occurring at V_{SET} and vice versa at V_{RESET}). It is evident that such stochasticity jeopardizes the operation of memory *e.g.* large HRS variation can result in erroneous read-out by the sense amplifier of the memory array. While ReRAM variations have been utilized for some applications like stochastic learning and physical unclonable functions [32, 33, 34], they remain a hurdle for memory and other deterministic computing applications.

In the past, researchers attempted to eradicate variability in resistive switching behavior by device engineering, *i.e.* by manipulating the switching oxide used and its composition. However, past efforts in device engineering seem to suggest that variability can be reduced, but not completely eliminated. There is a certain amount of variability which is intrinsic to Resistive RAM, *i.e.* variability is due to the stochastic nature of formation and rupture of the conductive filament [29, 28, 17]. The subject of variability and its causes is still a matter of intense research and is not completely understood. It is believed that the formation/rupture of the conductive filament is a stochastic process, by nature [28, 29, 17]. This stochastic process naturally results in stochasticity in the programmed resistance. It is reported that the cause of variability at HRS is due to variation of the number of particles in the narrowest current-controlling part of the filament [35], while the cause of variability at LRS is due to variation in the morphology of the filament, *i.e.*, its shape varies from cycle-to-cycle [36]. Origin of device-to-device variability is attributed to discrepancies in the fabrication processes such as variation in the switching oxide thickness, surface roughness of the electrodes, etching damages, etc., as well as the lack of precise control over the defect generation and filament formation during the 'forming' step of a pristine device [28].

2.6.1 How is variability Measured and Reported?

Variability in V_{SET}/ V_{RESET} is quantified as $\sigma_{SET}/\sigma_{RESET}$, the standard deviation from the mean, μ_{SET}/μ_{RESET}. Variability in HRS/LRS is also expressed as standard deviation, $\sigma_{HRS}/\sigma_{LRS}$ from the mean, μ_{HRS}/μ_{LRS}. While the V_{SET} and V_{RESET} voltages of ReRAMs are almost of the same order, the HRS can be up to three orders of magnitude higher than LRS. Therefore, the resistance variability is normalised by mean resistance and quantified by Co-efficient of Variation (CV), σ_{LRS}/μ_{LRS} and σ_{HRS}/μ_{HRS}. For example, the reported cycle-to-cycle variability for a Al/Ge/TaOx/Pt device is: σ_{SET} = 0.48 V, σ_{RESET} = 0.25 V, $(\sigma/\mu)_{LRS}$ = 25%, $(\sigma/\mu)_{HRS}$ = 80%. The device-to-device variability across the array (for the same device) is σ_{SET} = 0.39 V, σ_{RESET} = 0.29 V, $(\sigma/\mu)_{LRS}$ = 5%, $(\sigma/\mu)_{HRS}$ = 26% [37]. Clearly, the cycle-to-cycle variability is greater than device-to-device variability for both HRS and LRS for this device. Furthermore, many studies have revealed that variability is larger at HRS than at LRS [28, 38, 39] due to stochastic nature of the filament rupture. Few studies have compared the actual variation in HRS and LRS. For *e.g.*, in a HfO_x device, the reported variation is 16% at LRS and 36% at HRS. For TiO_x device, it is 14% at LRS and 26% at HRS [30].

2.6.2 How is Variability Controlled?

Researchers are pursuing different techniques to combat variability in ReRAM. Cycle-to-cycle variability can be tackled by program-and-verify algorithms [40]. In this algorithm, instead of writing using a single pulse, several progressively increasing pulses are used with READ pulses in between. The WRITE pulses are applied till a a target resistance is reached (the READ pulses in between WRITE pulses are used to monitor the current resistance). The goal is to have better control over the structure of the conductive filament formed, thereby minimizing cycle-to-cycle variability. This method comes at the cost of increased WRITE latency and programming energy. Device-to-device variability can be improved by targeting better uniformity across the wafer for the different integration steps [17].

Another technique to combat variability is called stack engineering [41]. In this technique, multiple layers of switching-oxides are used instead of a single switching-oxide layer between the electrodes. The purpose is to confine as much as possible all stochastic mechanisms by splitting the switching process through several layers. Examples include the introduction of an additional Al_2O_3 layer [42], germanium layer [37], $TiOx$ layer [43]. In the past decade, bidimensional materials (also known as 2D materials) like hexagonal boron nitride (BNO_x) have also been experimented and they claim less variability [44]. However, these 2D materials require a sophisticated fabrication process as compared to traditional transition metal oxides [41].

References

[1] L. Chua, "Memristor-the missing circuit element," *IEEE Transactions on Circuit Theory*, vol. 18, no. 5, pp. 507–519, 1971.

[2] L. Chua, "Resistance switching memories are memristors," in *Handbook of Memristor Networks* (L. Chua, G. C. Sirakoulis, and A. Adamatzky, eds.), pp. 197–230, Cham: Springer International Publishing, 2019.

[3] D. J. Wouters, R. Waser, and M. Wuttig, "Phase-change and redox-based resistive switching memories," *Proceedings of the IEEE*, vol. 103, no. 8, pp. 1274–1288, 2015.

[4] L. Goux, "Oxram technology development and performances," in *Advances in Non-Volatile Memory and Storage Technology (Second Edition)* (B. Magyari-Köpe and Y. Nishi, eds.), Woodhead Publishing Series in Electronic and Optical Materials, pp. 3–33, Woodhead Publishing, second edition ed., 2019.

[5] Y. Chen, "Reram: History, status, and future," *IEEE Transactions on Electron Devices*, vol. 67, no. 4, pp. 1420–1433, 2020.

[6] Z. Shen, C. Zhao, Y. Qi, W. Xu, Y. Liu, I. Z. Mitrovic, L. Yang, and C. Zhao, "Advances of rram devices: Resistive switching mechanisms, materials and bionic synaptic application," *Nanomaterials*, vol. 10, no. 8, 2020.

[7] E. Ambrosi, A. Bricalli, M. Laudato, and D. Ielmini, "Impact of oxide and electrode materials on the switching characteristics of oxide reram devices," *Faraday Discuss.*, vol. 213, pp. 87–98, 2019.

[8] S. Siegel, C. Baeumer, A. Gutsche, M. von Witzleben, R. Waser, S. Menzel, and R. Dittmann, "Trade-off between data retention and switching speed

in resistive switching reram devices," *Advanced Electronic Materials*, vol. 7, no. 1, p. 2000815, 2021.

[9] D. Ielmini, "Resistive switching memories based on metal oxides: mechanisms, reliability and scaling," *Semiconductor Science and Technology*, vol. 31, p. 063002, may 2016.

[10] A. J. Kenyon, A. Mehonic, W. Ng, L. Zhao, H. Cox, M. Buckwell, K. Patel, A. P. Knights, D. J. Mannion, and A. L. Shluger, "Defining the performance of siox reram by engineering oxide microstructure," in *2022 11th International Conference on Modern Circuits and Systems Technologies (MOCAST)*, pp. 1–4, 2022.

[11] S. Menzel, U. Böttger, M. Wimmer, and M. Salinga, "Physics of the switching kinetics in resistive memories," *Advanced Functional Materials*, vol. 25, no. 40, pp. 6306–6325, 2015.

[12] S. Gao, F. Zeng, F. Li, M. Wang, H. Mao, G. Wang, C. Song, and F. Pan, "Forming-free and self-rectifying resistive switching of the simple pt/taox/n-si structure for access device-free high-density memory application," *Nanoscale*, vol. 7, pp. 6031–6038, 2015.

[13] K. Vishwakarma, R. Kishore, and A. Datta, "Formation of remote quantum point contact inside dielectric medium of an alox/siox reram," *IEEE Transactions on Electron Devices*, vol. 69, no. 12, pp. 6738–6744, 2022.

[14] Z. Wan, H. Mohammad, Y. Zhao, R. B. Darling, and M. P. Anantram, "Bipolar resistive switching characteristics of thermally evaporated v2o5 thin films," *IEEE Electron Device Letters*, vol. 39, no. 9, pp. 1290–1293, 2018.

[15] U. Böttger, M. von Witzleben, V. Havel, K. Fleck, V. Rana, R. Waser, and S. Menzel, "Picosecond multilevel resistive switching in tantalum oxide thin films," *Scientific Reports*, vol. 10, 2020.

[16] W. Chen, S. Tappertzhofen, H. J. Barnaby, and M. N. Kozicki, "Sio2 based conductive bridging random access memory," *Journal of Electroceramics*, vol. 39, pp. 109–131, 2017.

[17] M. Lanza, H.-S. P. Wong, E. Pop, D. Ielmini, D. Strukov, B. C. Regan, L. Larcher, M. A. Villena, J. J. Yang, L. Goux, A. Belmonte, Y. Yang, F. M. Puglisi, J. Kang, B. Magyari-Köpe, E. Yalon, A. Kenyon, M. Buckwell, A. Mehonic, A. Shluger, H. Li, T.-H. Hou, B. Hudec, D. Akinwande, R. Ge, S. Ambrogio, J. B. Roldan, E. Miranda, J. Suñe, K. L. Pey, X. Wu, N. Raghavan, E. Wu, W. D. Lu, G. Navarro, W. Zhang, H. Wu, R. Li, A. Holleitner, U. Wurstbauer, M. C. Lemme, M. Liu, S. Long, Q. Liu, H. Lv, A. Padovani, P. Pavan, I. Valov, X. Jing, T. Han, K. Zhu, S. Chen, F. Hui, and Y. Shi, "Recommended methods to study resistive switching devices," *Advanced Electronic Materials*, vol. 5, no. 1, p. 1800143, 2019.

[18] D. Ielmini and H.-S. P. Wong, "In-memory computing with resistive switching devices," *Nature Electronics*, vol. 1, pp. 333 – 343, 2018.

[19] T. Schenk, M. Pešić, S. Slesazeck, U. Schroeder, and T. Mikolajick, "Memory technology—a primer for material scientists," *Reports on Progress in Physics*, vol. 83, p. 086501, jun 2020.

[20] P. Huang, B. Chen, Y. J. Wang, F. F. Zhang, L. Shen, R. Liu, L. Zeng, G. Du, X. Zhang, B. Gao, J. F. Kang, X. Y. Liu, X. P. Wang, B. B. Weng, Y. Z. Tang, G.-Q. Lo, and D.-L. Kwong, "Analytic model of endurance degradation and its practical applications for operation scheme optimization in metal

oxide based rram," in *2013 IEEE International Electron Devices Meeting*, pp. 22.5.1–22.5.4, 2013.

[21] H.-S. P. Wong, H.-Y. Lee, S. Yu, Y.-S. Chen, Y. Wu, P.-S. Chen, B. Lee, F. T. Chen, and M.-J. Tsai, "Metal–oxide rram," *Proceedings of the IEEE*, vol. 100, no. 6, pp. 1951–1970, 2012.

[22] D. Alfaro Robayo, G. Sassine, Q. Rafhay, G. Ghibaudo, G. Molas, and E. Nowak, "Endurance statistical behavior of resistive memories based on experimental and theoretical investigation," *IEEE Transactions on Electron Devices*, vol. 66, no. 8, pp. 3318–3325, 2019.

[23] Y. Y. Chen, B. Govoreanu, L. Goux, R. Degraeve, A. Fantini, G. S. Kar, D. J. Wouters, G. Groeseneken, J. A. Kittl, M. Jurczak, and L. Altimime, "Balancing set/reset pulse for greater than 10e10 endurance hfo2/hf 1t1r bipolar rram," *IEEE Transactions on Electron Devices*, vol. 59, no. 12, pp. 3243–3249, 2012.

[24] M. Lanza, R. Waser, D. Ielmini, J. J. Yang, L. Goux, J. Suñe, A. J. Kenyon, A. Mehonic, S. Spiga, V. Rana, S. Wiefels, S. Menzel, I. Valov, M. A. Villena, E. Miranda, X. Jing, F. Campabadal, M. B. Gonzalez, F. Aguirre, F. Palumbo, K. Zhu, J. B. Roldan, F. M. Puglisi, L. Larcher, T.-H. Hou, T. Prodromakis, Y. Yang, P. Huang, T. Wan, Y. Chai, K. L. Pey, N. Raghavan, S. Dueñas, T. Wang, Q. Xia, and S. Pazos, "Standards for the characterization of endurance in resistive switching devices," *ACS Nano*, vol. 15, no. 11, pp. 17214–17231, 2021. PMID: 34730935.

[25] M. Si, H.-Y. Cheng, T. Ando, G. Hu, and P. Ye, "Overview and outlook of emerging non-volatile memories," *MRS Bulletin*, vol. 46, 11 2021.

[26] G. Molas and E. Nowak, "Advances in emerging memory technologies: From data storage to artificial intelligence," *Applied Sciences*, vol. 11, no. 23, 2021.

[27] G. C. Adam, A. Khiat, and T. Prodromakis, "Challenges hindering memristive neuromorphic hardware from going mainstream," *Nature Communications*, vol. 9, 2018. doi: 10.1038/s41467-018-07565-4.

[28] A. Prakash and H. Hwang, "Multilevel cell storage and resistance variability in resistive random access memory," *Physical Sciences Reviews*, vol. 1, no. 6, pp. –, 2016.

[29] A. Fantini, L. Goux, R. Degraeve, D. J. Wouters, N. Raghavan, G. Kar, A. Belmonte, Y. . Chen, B. Govoreanu, and M. Jurczak, "Intrinsic switching variability in hfo2 rram," in *2013 5th IEEE International Memory Workshop*, pp. 30–33, May 2013.

[30] V. Parmar and M. Suri, "Exploiting variability in resistive memory devices for cognitive systems," in *Advances in Neuromorphic Hardware Exploiting Emerging Nanoscale Devices* (M. Suri, ed.), pp. 175–195, New Delhi: Springer India, 2017. doi: 10.1007/978-81-322-3703-7_9.

[31] Y.-F. Kao, W. C. Zhuang, C.-J. Lin, and Y.-C. King, "A study of the variability in contact resistive random access memory by stochastic vacancy model," *Nanoscale Research Letters*, vol. 13, p. 213, Jul 2018.

[32] S. Yu, B. Gao, Z. Fang, H. Yu, J. Kang, and H.-S. P. Wong, "Stochastic learning in oxide binary synaptic device for neuromorphic computing," *Frontiers in Neuroscience*, vol. 7, p. 186, 2013.

[33] W. Wang, Y. Li, M. Wang, L. Wang, Q. Liu, W. Banerjee, L. Li, and M. Liu, "A hardware neural network for handwritten digits recognition using binary

rram as synaptic weight element," in *2016 IEEE Silicon Nanoelectronics Workshop (SNW)*, pp. 50–51, 2016.

[34] A. Chen, "Utilizing the variability of resistive random access memory to implement reconfigurable physical unclonable functions," *IEEE Electron Device Letters*, vol. 36, no. 2, pp. 138–140, 2015.

[35] R. Degraeve, A. Fantini, N. Raghavan, L. Goux, S. Clima, B. Govoreanu, A. Belmonte, D. Linten, and M. Jurczak, "Causes and consequences of the stochastic aspect of filamentary rram," *Microelectronic Engineering*, vol. 147, pp. 171 – 175, 2015. Insulating Films on Semiconductors 2015.

[36] D. Ielmini and V. Milo, "Physics-based modeling approaches of resistive switching devices for memory and in-memory computing applications," *J. Comput. Electron.*, vol. 16, pp. 1121–1143, Dec. 2017.

[37] V. Y. Zhuo, Y. Jiang, R. Zhao, L. P. Shi, Y. Yang, T. C. Chong, and J. Robertson, "Improved switching uniformity and low-voltage operation in taox-based rram using ge reactive layer," *IEEE Electron Device Letters*, vol. 34, pp. 1130–1132, Sept 2013.

[38] E. Ambrosi, A. Bricalli, M. Laudato, and D. Ielmini, "Impact of oxide and electrode materials on the switching characteristics of oxide reram devices," *Faraday Discuss.*, pp. –, 2018.

[39] G. C. Adam, A. Khiat, and T. Prodromakis, "Challenges hindering memristive neuromorphic hardware from going mainstream," *Nature Communications*, vol. 9, 2018.

[40] E. Pérez, C. Zambelli, M. K. Mahadevaiah, P. Olivo, and C. Wenger, "Toward reliable multi-level operation in rram arrays: Improving post-algorithm stability and assessing endurance/data retention," *IEEE Journal of the Electron Devices Society*, vol. 7, pp. 740–747, 2019.

[41] J. B. Roldán, E. Miranda, D. Maldonado, A. N. Mikhaylov, N. V. Agudov, A. A. Dubkov, M. N. Koryazhkina, M. B. González, M. A. Villena, S. Poblador, M. Saludes-Tapia, R. Picos, F. Jiménez-Molinos, S. G. Stavrinides, E. Salvador, F. J. Alonso, F. Campabadal, B. Spagnolo, M. Lanza, and L. O. Chua, "Variability in resistive memories," *Advanced Intelligent Systems*, vol. n/a, no. n/a, p. 2200338.

[42] E. Shahrabi, C. Giovinazzo, J. Sandrini, and Y. Leblebici, "The key impact of incorporated al2o3 barrier layer on w-based reram switching performance," in *2018 14th Conference on Ph.D. Research in Microelectronics and Electronics (PRIME)*, pp. 69–72, July 2018.

[43] A. Hardtdegen, C. La Torre, F. Cüppers, S. Menzel, R. Waser, and S. Hoffmann-Eifert, "Improved switching stability and the effect of an internal series resistor in hfo2/tioxbilayer reram cells," *IEEE Transactions on Electron Devices*, vol. 65, pp. 3229–3236, Aug 2018.

[44] I. Sanchez Esqueda, H. Zhao, and H. Wang, "Efficient learning and crossbar operations with atomically-thin 2-D material compound synapses," *Journal of Applied Physics*, vol. 124, 10 2018. 152133.

Chapter 3
ReRAM Modeling

3.1 What is a Model for a Memory Device?

Roughly speaking, **modeling** refers to the way in which the physical switching phenomenon is imitated by a computer (software program) so that circuit designers can understand the working of a device and other larger circuits (using those devices) without fabricating the device. Because fabricating the memory array and the circuits around it (to read and write) is costly, researchers depend on models to get an initial feel or a feasibility prediction of a memory technology. This was the same case with CMOS technology and SPICE models were developed to design and simulate circuits to study the feasibility of a particular chip or microprocessor before going for large scale production. In the same way, once a memory technology attains reasonable maturity, models are developed to study the feasibility of a memory technology for commercial applications. Consider a 256×256 array of memory cells along with peripheral circuits to write and read from the memory array. If a model for a ReRAM device is developed, the memory array can be simulated. With SPICE models for CMOS peripheral circuits, reading and writing to the memory array can be verified by simulation. In this manner, the feasibility of the entire memory system can be studied (WRITE energy consumption, READ latency etc). This will aid in the progress and commercialization of a particular memory technology. Therefore, it is important for engineers/researchers to understand how a memory device is modeled so that circuit simulations can be performed using the models.

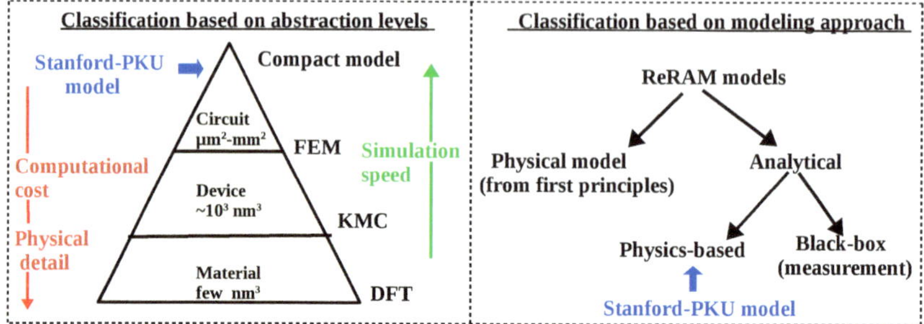

Fig. 3.1 Based on degree of detail, a model for ReRAM can range from atomistic (Density Functional Theory (DFT)) to Kinetic Monte Carlo (KMC) to Finite Element Method (FEM) to compact model [1]. From another perspective, a model for ReRAM can be derived from first principles or simplifications.

3.2 Approaches to ReRAM Device Modeling

The resistive switching phenomenon (the change in resistance of the ReRAM cell in response to voltage stress) has been modeled by various groups using different approaches [2, 1]. A classification of ReRAM models based on the degree of detail is presented in [1] (Fig. 3.1). While atomistic models capture intricate details like ion/atom diffusion and migration mechanisms at atomistic scale (few nm^3), compact models capture macroscopic details like geometry of the conductive filament and temperature. Based on the modeling approach, they can be classified as physical models and analytical models (Fig. 3.1). Physical models try to model resistive switching behavior from first principles, *i.e.* by modeling the fundamental cause for change in resistance, from the perspective of device-physics. On the other hand, analytical models are formulated as mathematical equations, which match the device's experimentally-observed behaviour. The physical principles approach and analytical approach are two extremes, and in reality, a model for ReRAM is a compromise between these two extremes [3]. Analytical approaches can be further classified as being either physics-based or black-box. In a physics-based analytical model, the physics of resistive switching is modeled by appropriate equations *e.g.* a physics-based analytical model may simplify resistive switching to the formation and rupture of conductive filament under voltage stress. In a black-box model, a measurement approach is followed and the approach is agnostic to the device structure or switching mechanism *e.g.* a device is subjected to different stimulus and its response is studied; then mathematical equations are formulated which obey this behavior (VTEAM model [4] is a classic example of black-box approach).

In addition to the aforementioned classification schemes, some models are specifically proposed for a ReRAM switching mechanism. Examples of these would be [5] and [6] that specifically model Valence Change Mechanism (VCM) and Electrochemical Metallization (ECM) devices, respectively. Physics-based models tend to be more accurate than black-box models since they consider temperature related phenomenon like joule heating, which influence the resistance to which the ReRAM is programmed during SET/RESET operation. Moreover, certain physics-based models not only capture switching physics (*i.e.* SET/RESET operation), but also certain temporal characteristics like reliability and technology related characteristic like scaling [1]. From the perspective of modeling detail (abstraction), the more detailed a physical model is, the higher the computational cost/simulation time. The simulation time of a ReRAM model increases drastically from 'Compact' models to 'Atomistic' models and hence accuracy and simulation time must be judiciously balanced. For circuit designers dealing with large number of devices in their circuit (*e.g.* a memory array), it is important to choose a compact model since the simulation speed is all the more important when numerous devices are involved. Around 15-20 distinct models for ReRAM device have been reported in literature. Our purpose of discussing modeling in this chapter is not to survey all the models proposed in recent literature. For the reader interested in review of ReRAM models, excellent surveys can be found in [1, 2, 3, 7, 8, 9, 10].

Since the focus of this book is memory circuit design, we are interested in knowing how to use a ReRAM device model to simulate a memory array. Therefore, we need to focus on a compact model for ReRAM. Furthermore, the circuit designer must be able to quickly fit the model (tune the model parameters) to different switching behaviours (the switching characteristics of ReRAM device depends on electrodes, switching material, its thickness, *etc.* and varies from fab to fab). Therefore, from the available ReRAM models, we choose the Stanford-PKU model and demonstrate how it can be used to mimic the resistive switching behaviour of different ReRAMs. The Stanford-PKU model is known for its 'structural stability' [11, 12]. If a model has structural stability, the qualitative properties of the model do not change in response to small changes in model parameters [13]. In other words, the model parameters can be varied (rather tuned) without affecting the key qualitative property of the model *i.e.* resistive switching behavior. This enables the Stanford-PKU model to be able to reproduce the resistive switching behaviour of a wide variety of ReRAMs which differ greatly in their switching characteristics.

3.3 Stanford-PKU ReRAM Model

The Stanford-PKU model is a physics-based, compact model developed for metal oxide bipolar ReRAMs [14, 15, 16, 11]. The model was well characterized on HfO_2 and HfO_x/TiO_x bilayer devices [11]. The resistive switching behaviour is modeled by the growth (during SET) and rupture (during RESET) of the conductive filament. Since filamentary switching is believed to the switching phenomenon in both oxide-based ReRAM and conductive bridging ReRAM [1, 8], the model is generic enough to model a wide variety of ReRAMs. This model simplifies the resistive switching behaviour into the growth and rupture of a single dominant filament. The gap distance, g (between the tip of the filament and the counter electrode) is the crucial parameter which determines the resistive state (see Fig. 3.2). The parameter g is programmable between gap_{min} and gap_{max} with the device being in HRS at gap_{max} and in LRS at gap_{min} (default values of gap_{min} and gap_{max} are 0.2 nm and 1.8 nm). Fig. 3.2 lists the equations governing the resistive switching process. The key equation (shaded yellow) describes the current (I) through the ReRAM as a function of voltage (V) across it and the gap in the conductive filament (g). The current has an exponential dependence on g, which together with hyperbolic dependence on the V, implements the sudden increase (or decrease) of the current resulting in a transition to LRS (or HRS). The reader is referred to [11] for an elaborate description of the model.

The parameters E_a (activation energy), a_0 (atomic spacing of the switching oxide), t_{ox} (thickness of the switching oxide), T_0 (environment temperature) and R_{TH} (thermal resistance) are determined by device structure, material properties and test environment. They can be easily obtained from the ReRAM manufacturer since they are process parameters. δ_g^0 is the fitting parameter for variations in the gap. T_{CRIT} denotes the threshold temperature, above which significant variations in the gap occurs and T_{SMTH} is the variations smoothing parameter [11]. These three parameters (δ_g^0, T_{CRIT}, T_{SMTH}) are related to variations in switching phenomenon and are used to model cycle-to-cycle variability in a ReRAM cell. The parameters – I_0, g_0, V_0, υ_0, γ_0 and β are called 'switching parameters' by the model developers, and they determine the

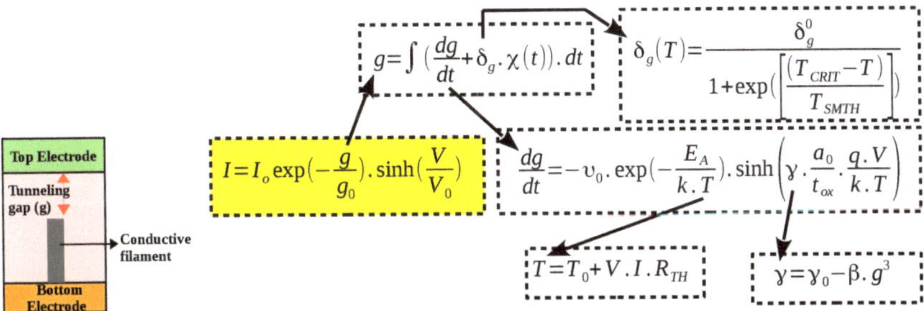

Fig. 3.2 Equations of Stanford-PKU ReRAM model (Redrawn from [17]): The tunneling gap 'g' is the state variable which models the resistance of the memory cell. Key equation (yellow) describes current through ReRAM as a function of the voltage across it and filament gap. The other equations describe the evolution of the gap 'g' and the associated temperature change [11].

median switching characteristics. The process parameters are dictated by the fabrication aspects of the device and are not tunable. Therefore, the six switching parameters are the key knobs to tune the model to mimic the characteristic of a particular ReRAM. An algorithm to tune these six switching parameters to fit the Stanford-PKU model to the characteristics of a particular ReRAM is presented in [17]. We elaborate on the algorithm so that the reader can use the algorithm to fit the Stanford-PKU model to reproduce a particular ReRAM's switching characteristics.

Table 3.1 A sample of fabricated ReRAMs with their median characteristics

Device	SET[V]	RESET[V]	LRS[Ω]	HRS[Ω]	Ref
$Pt/HfO_2/Ti/TiN$	0.88	-0.5	3.65K	5.1M	[18]
$TiN/Hf_{1-x}Al_xO_y/Ti/TiN$	0.9	-1.07	6.66K	66.66K	[19]
$Al/Ge/TaO_x/Pt$	2	-0.96	826	37M	[20]
$Ti/SiO_2/C$	2.4	-1.25	20K	100M	[21]

3.3.1 Fitting Algorithm

Different materials for the switching layer (metal oxide) and top/bottom electrodes have been explored by researchers to fabricate ReRAMs with different characteristics. In Table 3.1, four different ReRAMs and their median characteristics (switching voltages for SET/RESET and the LRS/HRS) are listed. The table is not comprehensive and represents a sample of ReRAMs devices from literature. The table highlights the diversity in ReRAM characteristics. One can easily observe that the RESET voltage varies from as low as -0.5 V to -1.25 V while the high resistance state (HRS) varies from 66.66 KΩ to as high as 100 MΩ, among a random sample of ReRAMs. At the same time, developing a ReRAM model is an onerous task involving many time-consuming experiments. Generally, research groups develop a model and verify their model on a device fabricated in their lab. After calibrating their model on measurements, the model is released/published with a default set of parameters which are fitted to that particular device (for example, the model proposed in [11] was verified on Al-doped HfOx devices and released with corresponding parameters which match their device's characteristics. Similarly, the 'cone' model in [22] was verified on $Ti/ZrO_2/Pt$ devices and released with fitting parameters). Often, there arises a need to be able to use a model for a device other than the device on which the model was verified. This needs a 'fitting algorithm' – an algorithm to tune the parameters of an established model to a particular ReRAM's switching characteristics. In this section, we will describe how the Stanford-PKU model can be tuned to fit a particular ReRAM's characteristics.

Table 3.2 The effect of 25% perturbation in the switching parameters of the Stanford-PKU ReRAM model

	Quantitative	Qualitative (Predominant role)
I_0	I_{HRS} and I_{LRS} increase by 25%	scales current uniformly
g_0	I_{HRS} and I_{LRS} increase by 420% and 20%, respectively	determines resistance window i.e. HRS/LRS ratio
V_0	I_{HRS} and I_{LRS} both decrease by 23%	determines slope of HRS or LRS
v_0	No change in current levels; V_{SET} and V_{RESET} decrease by 1.5 %	Since γ_0 has a stronger influence on V_{SET}, this parameter can be used to tune the voltage at which the device RESETs
γ_0	No change in current levels; V_{SET} decreases by 25%	determines the voltage at which the device SETs
β	No change in current levels; V_{SET} increases by 10% ; RESET process becomes more gradual	determines RESET curvature*

* the model implements gradual RESET while SET is abrupt and this is typical of ReRAM devices

By observing Fig.3.2, one can decipher that tuning the six switching parameters to fit the model to a ReRAM's specifications is a multi-faceted optimization problem. To investigate the correlation between the parameters and resistive switching behavior, small perturbations (of different degrees) were applied to each of the six parameters and its effect on the I-V curves were studied. When the parameters were perturbed simultaneously, the results were inconclusive and not useful. Therefore, a single parameter was perturbed (keeping other five constant) to find the predominant role of each parameter. The default values of each of these switching parameters (I_0 = $1e^{-3}$, g_0 = $2.5e^{-10}$, V_0 = 0.25, υ_0 = 10, γ_0 = 16, β = 0.8) was increased by 25%, one parameter at a time, and the resulting change in the I-V curves were observed. In Table 3.2, the vital effects of the perturbation in each of the six switching parameters are summarized. Except υ_0, which influences both V_{SET} and V_{RESET} equally, all the parameters have a specific role. Understanding this role gives insights into the parameter needed to be tuned to fit the model to a particular switching behavior. Based on this analysis, a fitting algorithm is proposed which efficiently fits the model to a specific ReRAM with minimum effort in terms of simulation iterations and time to fit [17].

Starting with the default parameters released with the Stanford-PKU model, the algorithm described in Fig. 3.3 tunes the different parameters of the model to fit to a target ReRAM. In simulation, the device must be set to an initial state, which is stable. This is implemented by the 'gap_{ini}' parameter which must be set to gap_{min} or gap_{max}. It is assumed that the ReRAM to be simulated is already 'formed' and hence the initial state is known. We fit the RESET process first, due to the lack of a parameter which strongly influences V_{RESET}. A negative pulse greater than the target device's V_{RESET} is applied with 'READ' pulses before and after the RESET process. The order of tuning is $g_0 \rightarrow I_0 \rightarrow V_0$ for the RESET process. This is because g_0 has the largest influence on the current levels, followed by I_0 and V_0. When tuning I_0, if the device reads out I_{LRS} at LRS, it will read out I_{HRS} at HRS since g_0 is already tuned. Since varying V_0 disturbs the I_{LRS} fixed in the previous step, I_0 is tuned again after V_0. Once the base current levels are tuned to match the measured I-V curves, we proceed to fit the voltage at which the device RESETs. It must be noted that υ_0 must be tuned to match the voltage at which the device **starts** to RESET and β must be tuned to match the gradual nature of the RESET process.

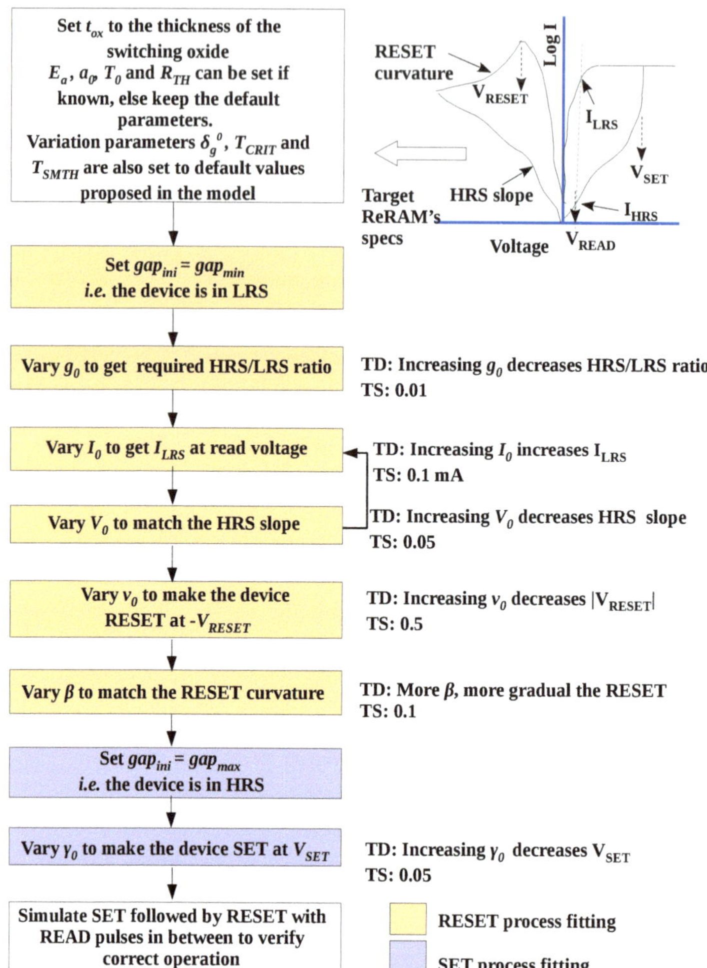

Fig. 3.3 Fitting Algorithm (Redrawn from [17]): In each step of the algorithm, 'TD' denotes 'Tuning Direction' to give direction to the tuning process and 'TS' denotes the 'Tuning Scale' to avoid unfeasible results

The fitting of the SET process is simpler and requires the tuning of only γ_0 to fix V_{SET} (the current levels during SET process are already tuned since I_0, g_0, V_0 affect both SET and RESET equally). Since γ_0 has no effect on the RESET process, the tuning of γ_0 towards the end does not 'disturb' the RESET curve already fitted. Since the RESET curvature and the slope of HRS/LRS are usually not quantified, V_0 and β are tuned till the simulated curves and measurement curves match sufficiently. This algorithm was used to fit the model to different ReRAMs we considered in Table 3.1. The simulated waveforms of some of the ReRAMs (from Table 3.1) during positive and negative voltages are graphed in Fig. 3.4. It was verified that the Stanford-PKU RRAM model was flexible enough to model ReRAMs of different materials (HfO_x, TaO_x, SiO_x) and different oxide thickness (5 nm $< t_{ox} <$ 10 nm). The tuned parameters are tabulated in Table 3.3 to make the curves reproducible. However, this fitting algorithm couldn't be used to fit ReRAMs with t_{ox} as high as 60 nm [23] and as low as 3 nm [24]. To fit those ReRAMs, the algorithm or the model itself may need to be significantly enhanced. Furthermore, this model was also used to reproduce cycle-to-cycle variability and device-to-device variability in simulations [25].

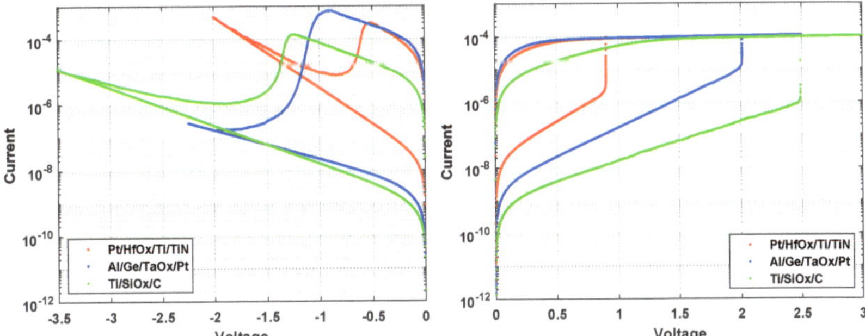

Fig. 3.4 Using the algorithm of Fig. 3.3, the model was fitted to ReRAMs with different SET/RESET voltages, HRS/LRS ratio, switching oxides and thickness (Redrawn from [17])

Table 3.3 The parameters of the Stanford-PKU model corresponding to different ReRAMs. Only t_{ox} and the parameters tuned in the fitting algorithm are tabulated, the remaining parameters are default values from the model release. δ_g^0=0.005 to minimize variations

Device	g_0	I_0	v_0	β	γ_0	V_0
$Pt/HfO_2(5nm)/Ti/TiN$	$2.176e^{-10}$	$0.17e^{-3}$	10.5	2.1	20.75	0.2
$Al/Ge/TaO_x(10nm)/Pt$	$1.495e^{-10}$	$1.04e^{-3}$	15	1.5	12.15	0.25
$Ti/SiO_2(5nm)/C$	$1.8525e^{-10}$	$0.374e^{-4}$	$1e^{-9}$	1.8	18	0.375

3.4 Steps to Simulate a ReRAM Device using Stanford-PKU Model in Cadence Virtuoso

Many researchers starting out in this field have a hard time with the first steps to simulate and perform simulation experiments using the ReRAM model. In this section, the basic steps to use the Stanford-PKU ReRAM model are summarized.

1. Download the Stanford-PKU ReRAM model, which is available for open-access from https://nanohub.org/publications/19/1. Please note that in the 'Supporting Docs' section of the above link, an accompanying manual is also provided by the model developers, which will be very helpful.
2. Create a symbol for ReRAM device having two terminals (top electrode and bottom electrode) in Cadence Virtuoso and link the Verilog-A code downloaded from *Nanohub* website (*rram_v_1_0_0.va*) to the ReRAM symbol.
3. Use the default model parameters and verify if the ReRAM device is working as intended. The default model parameters and the switching (SET/RESET) operations for those parameters are presented in the accompanying manual.
4. Use the set of model parameters provided in Table 3.4 and verify if the ReRAM device's switching characteristics changes.

In what follows, we will tune the Stanford-PKU model parameters to mimic a $Ti/SiO_2/C$ device having an oxide thickness of 5 nm. Right click on the ReRAM and go to 'Verilog-A properties'. There, all the model parameters and their default values will be available. Set the model parameters as depicted in Table 3.4 (*model_switch* parameter must be 0). To simulate the SET process, the device must first be in HRS. Set the 'gap_ini' parameter to 'gap_max' (this makes the device to be in HRS when the simulation starts). You can use a DC sweep from 0 to 3 V to check if the device undergoes a HRS to LRS transition at 2.4 V. Similarly, to simulate the RESET process, the device must first be in LRS. Set the 'gap_ini' parameter to 'gap_min'. Then, use a DC sweep of the voltage applied across the device from 0 to -2 V to see if the device resets at -1.25 volts. The multiple curves published for $Ti/SiOx/C$ device is depicted in Fig. 3.5 (left). We extracted the mean values and overlaid them on the model curve to highlight the fitting obtained in simulation (Fig. 3.5 (right)).

Table 3.4 Parameters of the Stanford-PKU model, their equivalent representation in the model equations (Fig. 3.2) and value to reproduce the switching characteristics of $Ti/SiO_2(5nm)/C$ device

Verilog-A parameter	Eq. representation(Fig. 3.2)	Value
kb	k	$1.38065e^{-23}$
q	q	$1.6e^{-19}$
g0	g_0	$1.8525e^{-10}$
V0	V_0	0.375
Vel0	v_0	$1e^{-9}$
I0	I_0	$0.374e^{-4}$
beta	β	1.8
gamma0	γ_0	18
T_crit	T_{CRIT}	450
deltaGap0	δ_g^0	0.02
T_smth	T_{SMTH}	500
Ea	E_a	0.6
a0	a_0	$2.5e^{-10}$
T_ini	-	298
F_min	-	$1.4e^{9}$
gap_ini	-	gap_min/gap_max
gap_min	-	$0.2e^{-9}$
gap_max	-	$1.8e^{-9}$
Rth	R_{TH}	2100
tox	t_{ox}	$5e^{-9}$
rand_seed_ini	-	0
time_step	-	$1e^{-9}$

References

[1] D. Ielmini and V. Milo, "Physics-based modeling approaches of resistive switching devices for memory and in-memory computing applications," *J. Comput. Electron.*, vol. 16, pp. 1121–1143, Dec. 2017.

[2] D. Panda, P. P. Sahu, and T. Y. Tseng, "A Collective Study on Modeling and Simulation of Resistive Random Access Memory," *Nanoscale Research Letters*, vol. 13, p. 8, Jan. 2018.

[3] R. S. Williams and M. D. Pickett, "The art and science of constructing a memristor model," in *Memristors and Memristive Systems* (R. Tetzlaff, ed.), pp. 93–104, New York, NY: Springer New York, 2014.

[4] S. Kvatinsky, M. Ramadan, E. G. Friedman, and A. Kolodny, "Vteam: A general model for voltage-controlled memristors," *IEEE Transactions on Circuits and Systems II: Express Briefs*, vol. 62, no. 8, pp. 786–790, 2015.

[5] C. La Torre, A. F. Zurhelle, T. Breuer, R. Waser, and S. Menzel, "Compact modeling of complementary switching in oxide-based reram devices," *IEEE*

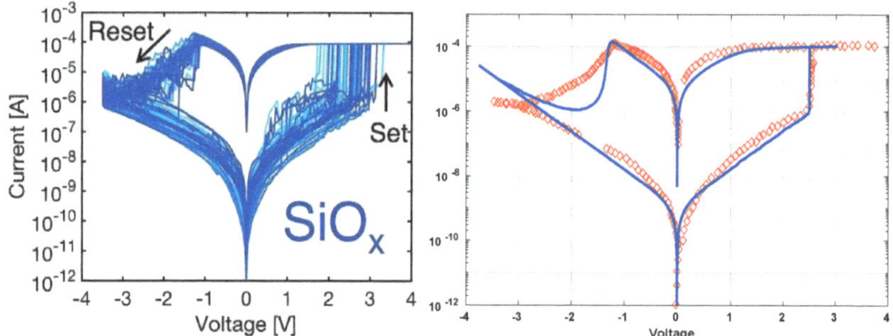

Fig. 3.5 Measured IV curves of $Ti/SiOx/C$ device for 50 cycles: Reproduced from Ref. [21] with permission from the Royal Society of Chemistry (Left). The Stanford-PKU model fitted to $Ti/SiOx/C$ device using the parameters in Table 3.4; red curve is the mean curve from measurements, blue curve is the IV characteristics produced by the model (Redrawn from [17],right).

Transactions on Electron Devices, vol. 66, pp. 1268–1275, March 2019.

[6] S. Menzel and R. Waser, "Analytical analysis of the generic set and reset characteristics of electrochemical metallization memory cells," *Nanoscale*, vol. 5, pp. 11003–11010, 2013.

[7] M. Lanza, H.-S. P. Wong, E. Pop, D. Ielmini, D. Strukov, B. C. Regan, L. Larcher, M. A. Villena, J. J. Yang, L. Goux, A. Belmonte, Y. Yang, F. M. Puglisi, J. Kang, B. Magyari-Köpe, E. Yalon, A. Kenyon, M. Buckwell, A. Mehonic, A. Shluger, H. Li, T.-H. Hou, B. Hudec, D. Akinwande, R. Ge, S. Ambrogio, J. B. Roldan, E. Miranda, J. Suñe, K. L. Pey, X. Wu, N. Raghavan, E. Wu, W. D. Lu, G. Navarro, W. Zhang, H. Wu, R. Li, A. Holleitner, U. Wurstbauer, M. C. Lemme, M. Liu, S. Long, Q. Liu, H. Lv, A. Padovani, P. Pavan, I. Valov, X. Jing, T. Han, K. Zhu, S. Chen, F. Hui, and Y. Shi, "Recommended methods to study resistive switching devices," *Advanced Electronic Materials*, vol. 5, no. 1, p. 1800143, 2019.

[8] S. Menzel, "Comprehensive modeling of electrochemical metallization memory cells," *Journal of Computational Electronics*, vol. 16, pp. 1017–1037, Dec 2017.

[9] E. Linn, A. Siemon, R. Waser, and S. Menzel, "Applicability of well-established memristive models for simulations of resistive switching devices," *IEEE Transactions on Circuits and Systems I: Regular Papers*, vol. 61, pp. 2402–2410, Aug 2014.

[10] B. Hajri, H. Aziza, M. M. Mansour, and A. Chehab, "Rram device models: A comparative analysis with experimental validation," *IEEE Access*, vol. 7, pp. 168963–168980, 2019.

[11] Z. Jiang, Y. Wu, S. Yu, L. Yang, K. Song, Z. Karim, and H. . P. Wong, "A compact model for metal–oxide resistive random access memory with experiment verification," *IEEE Transactions on Electron Devices*, vol. 63, pp. 1884–1892, May 2016.

[12] H. Li, Z. Jiang, P. Huang, Y. Wu, H. . Chen, B. Gao, X. Y. Liu, J. F. Kang, and H. . P. Wong, "Variation-aware, reliability-emphasized design and optimization of rram using spice model," in *2015 Design, Automation Test in Europe Conference Exhibition (DATE)*, pp. 1425–1430, March 2015.

[13] L. Chua, "Device modeling via nonlinear circuit elements," *IEEE Transactions on Circuits and Systems*, vol. 27, pp. 1014–1044, November 1980.

[14] X. Guan, S. Yu, and H. . P. Wong, "A spice compact model of metal oxide resistive switching memory with variations," *IEEE Electron Device Letters*, vol. 33, pp. 1405–1407, Oct 2012.

[15] S. Yu, B. Gao, Z. Fang, H. Yu, J. Kang, and H. . P. Wong, "A neuromorphic visual system using rram synaptic devices with sub-pj energy and tolerance to variability: Experimental characterization and large-scale modeling," in *2012 International Electron Devices Meeting*, pp. 10.4.1–10.4.4, Dec 2012.

[16] Z. Jiang, S. Yu, Y. Wu, J. H. Engel, X. Guan, and H. . P. Wong, "Verilog-a compact model for oxide-based resistive random access memory (rram)," in *2014 International Conference on Simulation of Semiconductor Processes and Devices (SISPAD)*, pp. 41–44, Sept 2014.

[17] J. Reuben, D. Fey, and C. Wenger, "A modeling methodology for resistive ram based on stanford-pku model with extended multilevel capability," *IEEE Transactions on Nanotechnology*, vol. 18, pp. 647–656, 2019.

[18] J. Sandrini, "Fabrication, characterization and integration of resistive random access memories," p. 241, 2017.

[19] E. Pérez, A. Grossi, C. Zambelli, P. Olivo, R. Roelofs, and C. Wenger, "Reduction of the cell-to-cell variability in hf1-xalxoy based rram arrays by using program algorithms," *IEEE Electron Device Letters*, vol. 38, no. 2, pp. 175–178, 2017.

[20] V. Y.-Q. Zhuo, Y. Jiang, R. Zhao, L. P. Shi, Y. Yang, T. C. Chong, and J. Robertson, "Improved switching uniformity and low-voltage operation in tao$_x$-based rram using ge reactive layer," *IEEE Electron Device Letters*, vol. 34, no. 9, pp. 1130–1132, 2013.

[21] E. Ambrosi, A. Bricalli, M. Laudato, and D. Ielmini, "Impact of oxide and electrode materials on the switching characteristics of oxide reram devices," *Faraday Discuss.*, vol. 213, pp. 87–98, 2019.

[22] G. González-Cordero, J. B. Roldan, F. Jiménez-Molinos, J. Suñé, S. Long, and M. Liu, "A new compact model for bipolar rrams based on truncated-cone conductive filaments—a verilog-a approach," *Semiconductor Science and Technology*, vol. 31, no. 11, p. 115013, 2016.

[23] Y. Huang, Z. Shen, Y. Wu, M. Xie, Y. Hu, S. Zhang, X. Shi, and H. Zeng, "Cuo/zno memristors via oxygen or metal migration controlled by electrodes," *AIP Advances*, vol. 6, no. 2, p. 025018, 2016.

[24] A. Bricalli, E. Ambrosi, M. Laudato, M. Maestro, R. Rodriguez, and D. Ielmini, "Resistive switching device technology based on silicon oxide for improved on–off ratio—part i: Memory devices," *IEEE Transactions on Electron Devices*, vol. 65, pp. 115–121, Jan 2018.

[25] J. Reuben, M. Biglari, and D. Fey, "Incorporating variability of resistive ram in circuit simulations using the stanford–pku model," *IEEE Transactions on Nanotechnology*, vol. 19, pp. 508–518, 2020.

Chapter 4
Array Structure and Memory Density

4.1 Memory Array

The non-volatile memory cells explored in the previous chapters have to be arranged in some fashion to be usable for storing and retrieving data in an organised manner. Conventional memories like SRAM and DRAM were always arranged in arrays. We define the memory array as a regular array of memory cells to store data, replicated in two dimensions– rows and columns. As depicted in Fig. 4.1 (a), the memory array of 8 rows and 8 columns can store 64 bits of information, where each memory cell stores a bit. We can call this the logical view. Just like a NAND gate is represented by a symbol in a circuit diagram but the detailed structure is made of 4 transistors (2 PMOS and 2 NMOS transistors), the structural view of the memory array is different from this logical view. The logical view is a more abstract view than the structural view. Therefore the logical view is the same but the structural view is different for different memory technologies. To illustrate, Fig. 4.1 (b) and (c) depict the detailed structure of 4×4 DRAM and SRAM array, respectively. We can call this the structural view. In practise, the structural view is important for circuit designers while those working higher at a system level will only 'see' the memory as arrangement of bits of information (in rows and columns).

As depicted, the structural view is dependent on the memory technology. For DRAM, each memory cell is a capacitor with an access transistor. But for SRAM, each memory cell is back-to-back connected inverters with two access transistors. Not only is the structural view different for different memory technologies, the way in which a bit of information is stored is also significantly different in different memory

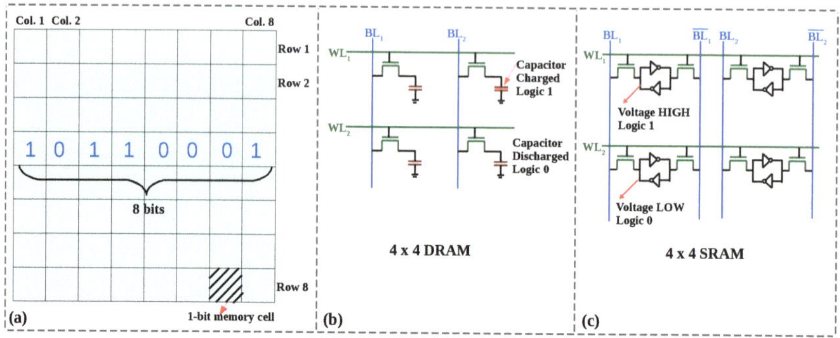

Fig. 4.1 (a) Logical view of the memory array: Each bit is just a location in the memory array. (b) Structural view of DRAM: Each bit of information is stored as a charge in the capacitor (c) Structural view of SRAM: Each bit is charge at the input of the back-to-back connected inverters

technologies. In DRAM, the information is stored as the charge in a capacitor while in SRAM the information is stored as a charge, which results in a voltage at the cross-coupled inverter. Another factor which is obvious from the structural view is the area occupied by a memory array in silicon. Clearly, 256×256 DRAM array will not occupy the same area as 256×256 SRAM array due to the difference in the area occupied by a single memory cell (Area of single DRAM cell is $6F^2$ while that of SRAM cell is $200F^2$ [1], where F is the feature size, explained in the next section). We have elaborated on the logical view and the structural view just to clarify the basic terminologies of the conventional memories. In the following sections, the commonly pursued array structures for ReRAMs will be presented.

4.2 Resistive RAM Array Configurations

The 1R configuration (1 **R**eRAM device forms a memory cell) is simply the ReRAM device fabricated between two metal layers. The **1S–1R** configuration is the same as 1R with a selector device fabricated in series with the ReRAM (1 **S**elector and 1 **R**eRAM device form a memory cell). **1T–1R** is the configuration where each memory cell is fabricated with a transistor to enable access to the memory device (1 **T**ransistor and 1 **R**eRAM device form a memory cell).

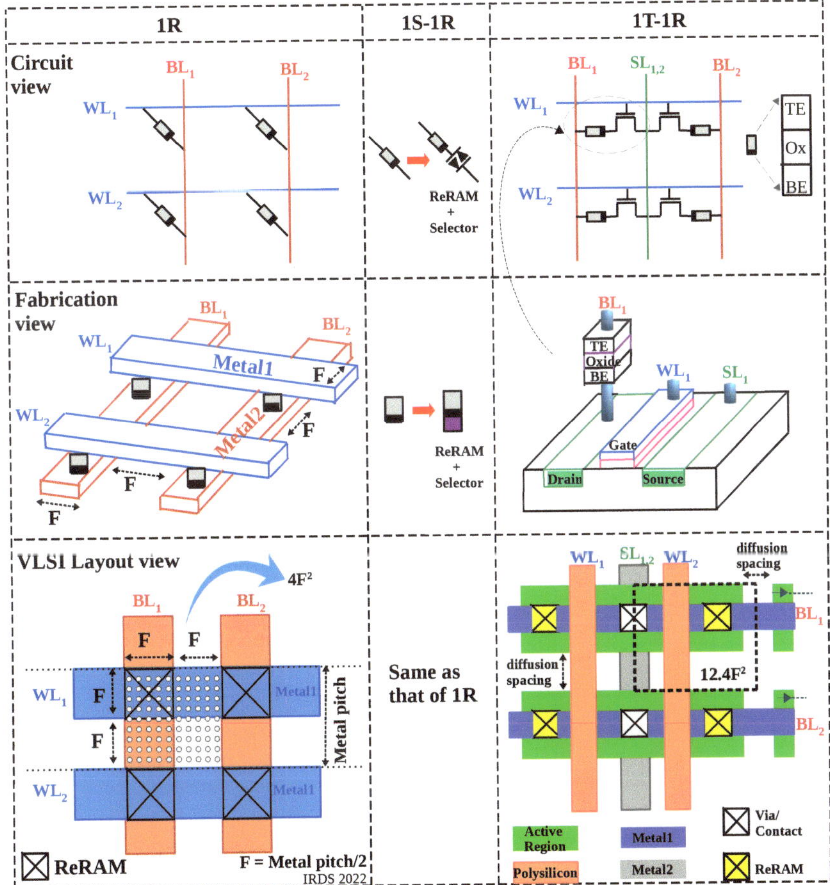

Fig. 4.2 Illustration of different array structures typically used to construct ReRAMs. 1S–1R is essentially the 1R configuration with the ReRAM device appended with a selector device to avoid sneak paths (The layout view of 1T-1R array is re-drawn from the layout presented in [2])

4.2.1 1R Configuration

First, we consider the 1R configuration. Here, the ReRAM is fabricated between two metal lines in a BEOL process*. As depicted in Fig. 4.2,

* Back-End-Of-Line process is the step in the VLSI fabrication process in which the interconnection between transistors are made using metals after fabricating the MOSFET's drain, gate, source and substrate in Front-End-of-Line (FEOL) process

the ReRAM device can be fabricated between two metal layers, which run perpendicular to each other. This structure is also called cross-point or crossbar since the ReRAM device is at the crossing of the two metal lines. It is a common practise to compare the array area of different memory technologies in terms of F^2, where F is the minimum feature size of a given lithographic technology [3] ($F = \frac{Metal-Pitch}{2}$, IRDS 2022 Executive summary†). Consequently, the area of a single memory cell in 1R configuration is $2F \times 2F = 4F^2$, as depicted in Fig. 4.2. Comparing layout area in terms of F^2 instead of nm^2 enable us to get a quick estimate of the memory area independent of the process node [1].

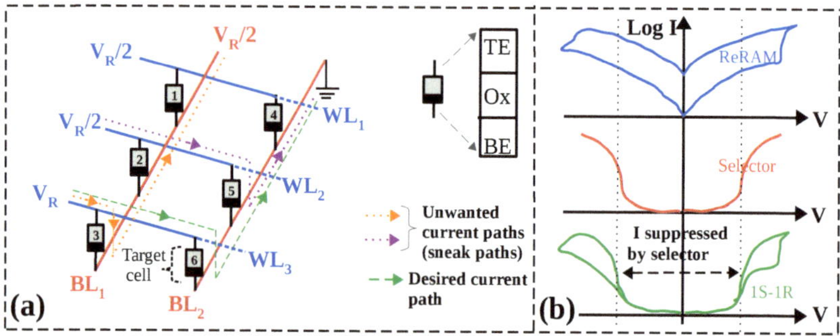

Fig. 4.3 (a) Illustration of sneak path currents during READ operation. During WRITE operation also, similar phenomenon occurs, making both reading and writing energy inefficient (b) Selector suppresses current through the 1S-1R cell at low voltages

4.2.2 1S–1R Configuration

The 1R (crossbar) array configuration, although area-efficient, comes with an inherent inter-cell interference *i.e.* the inability to manipulate a single memory cell without affecting its neighboring cells in the memory array.

† Definition of Feature size F has undergone some changes in the last 30 years. It was defined in 1992 as one half of the dimension of the densest Metal Pitch but was often identified with gate length in later years since the dimensions of the gate length and the half-pitch of tightest metal layer were almost the same. In lower technology nodes, this is not the case and we stick to the original definition, as clarified by IRDS 2022 Executive summary [4]

When reading or writing to a particular cell, this interference causes leakage currents called 'sneak-path' currents to flow through un-selected cells. Fig.4.3 (a) illustrates sneak paths during READ operation. Typically, in 1R configuration, a ReRAM cell is read by applying a read voltage V_R across the cell while other cells should ideally be biased at 0 V. This is accomplished by applying $\frac{V_R}{2}$ at other WLs and BLs which are not connected to the target cell. In this illustration, we intend to read ReRAM cell 6. Since cells 3 and 6 share WL_3, cell 3 has $\frac{V_R}{2}$ across it. As a result, there is an unwanted current from WL_3 to BL_1 through cell 3. A similar unwanted current flows from WL_2 to ground through cell 5 since it has a voltage $\frac{V_R}{2}$ across it. These unwanted currents are called sneak-path currents. Notice that cells 1 and 2 do not conduct any such current since they have 0 V across them.The sneak-path problem in 1S–1R memory arrays essentially makes the reading and writing process energy in-efficient [5]. In addition, sneak-path currents may cause read-out errors and also limit the maximum size of the array [6].

Due to the sneak-path phenomenon and disadvantages which come with it, ReRAM is typically not manufactured in 1R configuration. Researchers and engineers have addressed this problem in two ways – 1) by using a 'selector' device (1S–1R) or 2) by using a transistor (1T–1R). Both configurations enable us to select a single cell without interfering with neighboring cells, but to different degrees. The former suppresses the sneak paths, while the later completely eliminates it. A 'selector' is a two-terminal device inserted in series with the ReRAM cell to minimize the sneak-path currents. The selector is essentially a device with strong non-linearity in its I-V characteristics, like a p-n junction diode [7]. Therefore, it suppresses current through it at low voltages and gradually allows current at higher voltages, as depicted in Fig. 4.3 (b). When reading/writing to a particular cell, the WLs/BLs of other cells are biased at V/2 or V/3 [8]. Since a selector suppresses current at lower voltages, it suppresses sneak currents through the other cells biased at V/2 or V/3. It should be emphasized that the selector minimizes the sneak current, but does not eliminate it completely. Even with a selector, the interference with the neighboring cells exist since the composite 1S-1R cell is integrated between the metal lines, as in 1R configuration. However, by carefully choosing the READ/WRITE voltage, the neighboring cells of the target cell can be biased at voltages such that the sneak current through them is reduced by orders of magnitude. It must be emphasized that the 1S–1R structure is advantageous even with this partial suppressing (of the sneak currents) due to the $4F^2$ area it occupies. Practical demonstrations of ReRAM fabricated in 1S-1R configuration can be found in [9, 10].

4.2.3 1T–1R Configuration

As stated, the other alternative to overcome the sneak-path issue is to integrate the ReRAM device with an access transistor for each cell, as illustrated in Fig. 4.2. As depicted, the ReRAM is fabricated above the transistor at the drain node. The bottom electrode is connected to the NMOS transistor's drain while the top electrode is connected to the BL. The 1T–1R array depicted in Fig. 4.2 is common source line configuration *i.e.* the 1T-1R cells in the adjacent columns share the source. This is pursued to minimize the 1T–1R cell's area [2]. The detailed steps in the fabrication process of ReRAM above the transistor is presented in [11]. Transistor plays two important roles in 1T–1R configuration. Firstly, it acts as an access device. To read or write into a cell, it must first be 'accessed' by switching ON the transistor with an appropriate voltage at the gate/WL. This is because a current can never flow from BL to SL (through the ReRAM) if the NMOS transistor is OFF. Only when the NMOS transistor is ON, the ReRAM cell can be read or written into by applying appropriate voltages at BL and SL. Secondly, the NMOS transistor acts as a current-limiting device. During the SET process (when the device switches from HRS to LRS), there will be a sudden increase in current and the maximum current that can flow through the 1T-1R cell is controlled/limited by the drain current of the transistor [3, 11]. By applying appropriate voltage at the gate of the transistor, the current through the 1T-1R cell is controlled and this stabilizes the $V_{SET/RESET}$ and $I_{SET/RESET}$ [12, 13]. The switching current through the 1T-1R cell is also called compliance current since the current through ReRAM cell is made to comply with the current through transistor. In summary, integrating a transistor with the ReRAM makes the resistive switching process more controllable and reduces variability in the programmed resistance. However, this advantage comes at the cost of increased area (due to the transistor) when compared with 1R and 1S–1R configuration. With access transistor, area of a single 1T–1R memory cell is theoretically $8F^2$ [3]. Practically, the 1T–1R cell area varies from $15F^2$ [14] to $30F^2$ [15]. In spite of the area penalty, it must be emphasized that the 1T-1R configuration is the most reliable and practical array configuration for ReRAM arrays. ReRAM prototypes by National Tsing

Hua University and TSMC [16, 17], CEA/LETI [18] and the recent RAM chip with Compute-in-memory capability demonstrated by Stanford NVM Alliance [19] are all fabricated in 1T-1R configurations.

4.3 Selectors for 1S–1R Array

4.3.1 Required Characteristics

A selector is simply a device with strong non-linear characteristics, fabricated in series with a ReRAM between the WL and BL (Fig. 4.2). The design of a selector is non-trivial since the combination (ReRAM device + selector) must function as a single entity, suppressing the leakage current through unselected cells (*i.e.* it must restrict current through cells where READ/WRITE operation is not intended) and allowing the current through selected cell. The 1S-1R cell must 'know' whether it is selected or not based on the voltage across it. In other words, the 1S-1R cell must be carefully designed such that it restricts current at lower voltages and allows current at higher voltages. Burr et al. identified the required characteristics of a selector as follows [20]:

1. **High ON-state current density**: The selector must be capable of supplying enough current to program the cell it is integrated with. Since the cross-section of the cell is only $1F^2$ in 1S-1R array, the selector requires a high current density of the order of several Mega-ampere per square centimeter.
2. **Low OFF-state leakage current**: Since the unselected cells largely outnumber the selected cells, the OFF-state leakage current should be as low as possible in order to conserve the energy consumption during reading and writing.
3. **Bidirectional operation**: If the ReRAM has bipolar switching characteristics, the selector must also be capable of bipolar operation. In other words, it must be able to behave like a non-linear device not only when positive voltage polarity is applied across it but also when negative voltage polarity is applied across it.
4. **Fabrication Process/BEOL compatibility**: The fabrication process of the selector must be compatible with Back-End-Of-Line (BEOL) process in which the ReRAM is fabricated. The maximum thermal budget during selector fabrication must not exceed 400 deg C [20].

5. **Voltage compatibility with ReRAM**: The selector's I-V characteristic should align with the ReRAM device's I-V characteristics. As depicted in Fig. 4.3 (b), the selector must allow as much current a ReRAM device allows at its switching voltages so that the ReRAM device can switch.
6. **Switching speed, endurance and variability of the selector must be same or better than that of the ReRAM device**. In other words, the memory device's switching speed or endurance or variability which decides its performance must not be worsened by the addition of a selector device.

These requirements complicate the fabrication of ReRAM device with a selector very challenging.

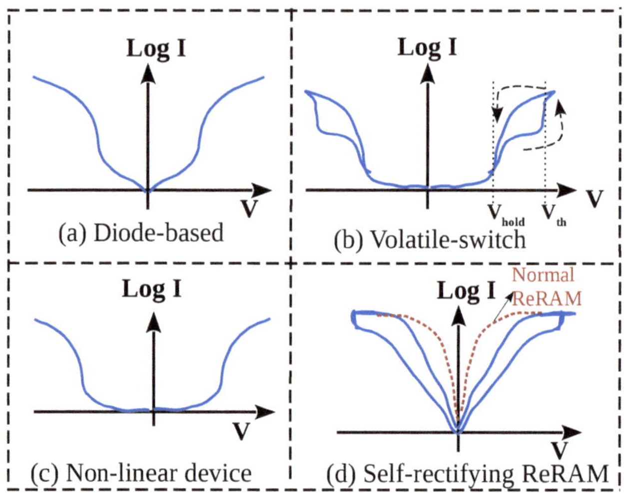

Fig. 4.4 I-V characteristics of two-terminal selectors for 1S-1R ReRAM array

4.3.2 Types of Selectors

Notwithstanding these difficult requirements, academia and industry have been relentlessly pursuing an ideal selector because of the area advantages the 1S-1R configuration offers. The major types of two-terminal selectors researched in the last decade are as follows [21]:

1. **Diode-based selectors**: Oxide Heterojunction diodes and Metal/Oxide Schottky diodes have been experimented and low processing temperature is one of their major advantages. They also have material compatibility with oxide-based ReRAMs.
2. **Volatile switches** are two-terminal selectors whose conductivity changes suddenly above a certain threshold voltage. For example, niobium oxide, vanadium oxide (also called Mott material) exhibit this property and have been experimented as selector devices with resistive memory devices.
3. **Non-linear devices**: Certain devices can be used as selectors because of the non-linearity they offer *i.e.* they become more resistive at lower voltages and become less resistive at higher voltages. These selectors exploit the tunneling-based transport mechanisms in oxide materials.
4. **Self-rectifying devices**: A tunneling-oxide layer intentionally built into a ReRAM device can provide non-linearity in the ReRAM's I-V characteristics. Such a device has an intrinsic 'selecting capability' and therefore called self-rectifying or self-selecting ReRAMs.

Fig. 4.4 plots the I-V characteristics of the different types of selectors. Diode-based selectors have the characteristics of a junction diode with bipolar characteristics. Volatile switches have a threshold voltage (V_{th}) beyond which the conductivity increases suddenly and a hold voltage (V_{hold}) below which the conductivity decreases drastically. The I-V characteristics of a self rectifying ReRAM device will be similar to that of a ReRAM device, except that the current is suppressed at lower voltages. Finally, the reader must be aware that when a transistor is integrated with the memory device, the transistor also functions as a control device to control the current through the memory device (by the gate voltage of the access transistor). The two-terminal selectors discussed in this section cannot provide this control. Therefore, this control is usually provided by peripheral transistors outside the 1S-1R array.

4.4 Three Ways to Increase Memory Density

Consumers want more and more data storage capacity on their portable devices. At the same time, nobody is willing to pay for it in terms of area. In other words, we want to increase memory density *i.e.* the number of bits that can be stored per unit area of a memory chip. To satisfy this requirement of more memory density, researchers have reacted in three ways – a) Device scaling b) Multi-Level Cell c) 3D stacking.

4.4.1 Device Scaling and Scalability

Scaling is important in memory technologies because the size of an individual memory cell dictates the memory density. The requirement for increased memory density has pushed memory designers to scale or shrink the size of the individual memory device. In the chapter on device characteristics, we should have discussed how the ReRAM device performs when scaled down *i.e.* when the size of the device shrinks. But we waited for this chapter to have a better context since the fabrication view and layout view of different array configurations are presented in detail in the beginning of this chapter. As elaborated, memory cells can be organised to construct arrays of different configurations, with 1S–1R and 1T-1R being the most common ones. A single ReRAM cell occupies an area of $4F^2$ when organised in 1S–1R configuration, with the ReRAM device of size $F \times F$. The question remains - will the ReRAM device continue to function as before even when F is scaled down? For example, if F is scaled down from 130 nm to 90 nm, the size of ReRAM cell will also reduce from 130 nm×130 nm to 90 nm×90 nm (Fig. 4.2). Scalability refers to the ability of a memory device to function as before even when the device size shrinks. ReRAM device exhibit good scalability and it is one of its significant strengths compared to competing memory technologies [22]. This is evidenced by many works where ReRAM devices with sub-100 nm feature size have been fabricated and they retained good performance in spite of the scaling, *e.g.* 30 nm×30 nm device [23] and 10 nm×10 nm device [24]. The reason behind ReRAM's good scalability is that the essential switching characteristics of the device are decided by the conductive filament, which can be as small as a few nanometers [25]. An interesting characteristic that has been observed is that while scaling, the HRS increases while LRS remains almost constant [26]. Consequently, the resistance window increases ($\frac{HRS}{LRS}$ ratio) and this is advantageous for the sensing operation.

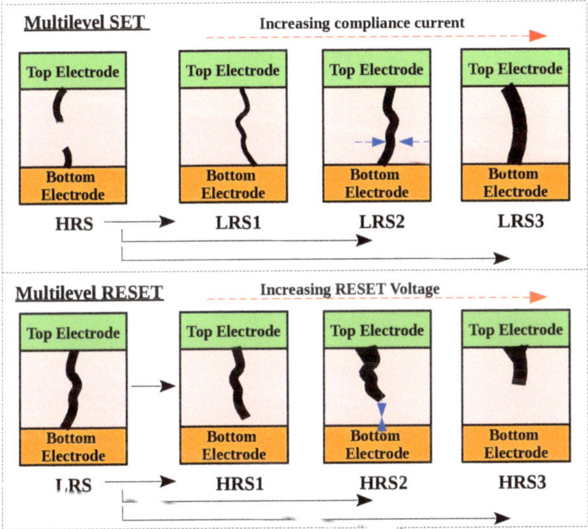

Fig. 4.5 Illustration of storing more than two distinct resistance. Varying the compliance current during SET process results in multiple LRS due to the varying thickness of the conductive filament. Varying the RESET voltage during RESET process results in multiple HRS due to the varying distance between the tip of the conductive filament and the bottom electrode.

4.4.2 Multi-Level Cell (Multi-bit)

To increase memory density, the second approach pursued was to store more than one bit per cell. In fact, this approach is not new and it was followed in the earlier flash memory technology. In flash memory, bit '1' and '0' were encoded as either no charge or charge in the floating gate which correspond to two different threshold voltages during read-out. This was called single-level cell. In Multi-Level Cell (MLC), the amount of charge placed in the floating gate was varied resulting in different threshold voltages. Conventionally, the ReRAM device could store 1 bit of information in a cell, exhibiting itself as HRS or LRS depending on whether the conductive filament was formed or ruptured. MLC can be achieved in ReRAM technology in the following ways [27]:

1. By varying compliance current (also called multilevel SET): from the initial HRS, the ReRAM cell is programmed to different LRS during the SET process.

2. By varying reset voltage (also called multilevel RESET): from the initial LRS, the ReRAM cell is programmed to different HRS during the RESET process.
3. By varying programming pulse widths (not popularly pursued due to being energy inefficient).

In 1T-1R arrays, multilevel SET is achieved by varying the gate voltage of the access transistor, which in turn varies the compliance (drain) current. During the SET process (HRS $\rightarrow$ LRS), the ReRAM cell gets programmed to three different LRS corresponding to three different compliance currents (determined by the access transistor's gate voltage). For example, the 1T–1R cell of IHP[‡] with a HRS of 66.6 KΩ can be programmed to LRS of 10KΩ, 6.6 KΩ and 5 KΩ by applying a gate voltage of 1.2 V, 1.4 V and 1.6 V, respectively [28]. The higher the compliance current, the lower the resistance the device is programmed to. The physical phenomenon behind this process is believed to be the formation and subsequent widening of the conductive filament with increasing compliance current [29, 30] (Fig.4.5). Multilevel RESET, which is implemented by varying the maximum voltage applied during RESET is well suited for passive arrays (1R and 1S-1R configuration). A HfOx-based ReRAM was programmed to three different HRS by applying a voltage of -0.9 V, -1.1 V and -1.3 V during RESET operation, while the LRS remained the same [27]. The physical phenomenon is believed to be the larger gap (between the tip of the conductive filament and bottom electrode) with increasing reset voltage, *i.e.* the device goes to a deeper RESET with higher reset voltage [27]. Finally, MLC can be implemented by varying the width of the programming pulse while its amplitude remains constant. In [31], three distinct HRS levels were obtained by varying the pulse width from 50 ns to 5 μs. Due to the high energy consumption during programming, this method is usually not favoured.

4.4.3 3D Stacking

Just like stacking housing units one over the other can increase the number of living units per km^2 of a city, stacking memory cells can increase the number of memory cells per mm^2 of a integrated chip. Again, stacking of ReRAM cells one over the other in a 3D fashion drew inspiration by the

[‡] Innovations for High Performance Microelectronics, Frankfurt Oder, Germany

success it had in flash memory technology. One must remember that the 3D *NAND* flash (also called as Bit-Cost Scalable flash [32]) had profound impact on the NVM market and sustained the popularity of flash memory technology for several years. When ReRAM technology reached sufficient maturity, it was natural for researchers to explore 3D stacking in the footsteps of flash memory technology.

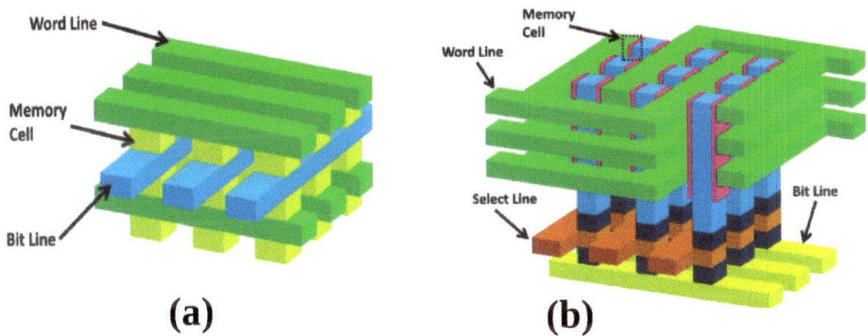

Fig. 4.6 Illustration of 3D stacking of ReRAM (a) Horizontally stacked 3D ReRAM (b) Vertically stacked 3D ReRAM. Reproduced, with permission, from [33]. Copyright 2013, IEEE.

4.4.3.1 Horizontally Stacked 3D ReRAM

The first method of 3D stacking was just placing layers of planar cross-bar arrays (1S–1R fabricated between two metal lines, Fig. 4.1) one over the other. As depicted in Fig. 4.6, the memory cells in a layer are placed between the wordline and bitline and the memory cells in the adjacent layer are placed above (or below) in a similar fashion. When they are placed so, the memory cells in adjacent layers share a electrode (blue bit lines are shared between the two layers of memory cells, Fig. 4.6 (a)). Theoretically, compared to planar array[§] where each cell occupies $4F^2$, for n horizontally stacked arrays, each cell occupies $4F^2/n$. However, the number of lithography steps, i.e., costs, rise proportionally with each stacked layer, limiting its 3D potential [34]. Intel/Micron's 3D Xpoint is an example of horizontally stacked 3D memory although Phase Change Memory (PCM) cells was used instead of ReRAM cells [35, 36].

§ single layer of ReRAM cells in 1S–1R configuration

4.4.3.2 Vertically Stacked 3D ReRAM

This structure is similar to the Bit-Cost Scalable 3D $NAND$ flash structure and is thus cost-effective for integrating many layers. As depicted in Fig. 4.6 (b), the memory cell is located between the green wordline and the vertical blue pillars, which act as bitline. Two sets of wordlines are interleaved. The blue vertical lines are connected to the yellow bitlines through transistors (shown in orange). The select lines which are connected to the gate of the transistors are used to select a vertical pillar, thus selecting a memory cell for reading and writing [37]. Vertically stacked 3D ReRAM is cost effective because critical patterning steps are performed once for all memory layers together [38]. The reader is referred to [39] for more recent developments in 3D stacking of ReRAM technology.

References

[1] T. Schenk, M. Pešić, S. Slesazeck, U. Schroeder, and T. Mikolajick, "Memory technology—a primer for material scientists," *Reports on Progress in Physics*, vol. 83, p. 086501, jun 2020.

[2] A. Levisse, P.-E. Gaillardon, B. Giraud, I. O'Connor, J.-P. Noel, M. Moreau, and J.-M. Portal, "Resistive switching memory architecture based on polarity controllable selectors," *IEEE Transactions on Nanotechnology*, vol. 18, pp. 183–194, 2019.

[3] D. Ielmini, "Resistive-switching memory," in *Wiley Encyclopedia of Electrical and Electronics Engineering*, pp. 1–32, John Wiley and Sons, Ltd, 2014.

[4] "International roadmap for devices and systems, 2022 executive summary. available:https://irds.ieee.org/editions/2022/executive-summary."

[5] A. Chen, "Memory selector devices and crossbar array design: a modeling-based assessment," *Journal of Computational Electronics*, Sep 2017.

[6] S. Kim, J. Zhou, and W. D. Lu, "Crossbar rram arrays: Selector device requirements during write operation," *IEEE Transactions on Electron Devices*, vol. 61, no. 8, pp. 2820–2826, 2014.

[7] G. W. Burr, R. S. Shenoy, K. Virwani, P. Narayanan, A. Padilla, B. Kurdi, and H. Hwang, "Access devices for 3d crosspoint memory," *Journal of Vacuum Science & Technology B, Nanotechnology and Microelectronics: Materials, Processing, Measurement, and Phenomena*, vol. 32, no. 4, p. 040802, 2014.

[8] S. Kim, H.-D. Kim, and S.-J. Choi, "Numerical study of read scheme in one-selector one-resistor crossbar array," *Solid-State Electronics*, vol. 114, no. Supplement C, pp. 80 – 86, 2015.

[9] T.-y. Liu, T. H. Yan, R. Scheuerlein, Y. Chen, J. K. Lee, G. Balakrishnan, G. Yee, H. Zhang, A. Yap, J. Ouyang, T. Sasaki, A. Al-Shamma, C. Chen, M. Gupta, G. Hilton, A. Kathuria, V. Lai, M. Matsumoto, A. Nigam, A. Pai, J. Pakhale, C. H. Siau, X. Wu, Y. Yin, N. Nagel, Y. Tanaka, M. Higashitani,

T. Minvielle, C. Gorla, T. Tsukamoto, T. Yamaguchi, M. Okajima, T. Okamura, S. Takase, H. Inoue, and L. Fasoli, "A 130.7-mm2 2-layer 32-gb reram memory device in 24-nm technology," *IEEE Journal of Solid-State Circuits*, vol. 49, no. 1, pp. 140–153, 2014.

[10] J. M. Lopez, D. Alfaro Robayo, L. Grenouillet, C. Carabasse, G. Navarro, R. Fournel, C. Sabbione, M. Bernard, O. Billoint, C. Cagli, L. Couture, D. Deleruyelle, M. Bocquet, J. M. Portal, E. Nowak, and G. Molas, "Optimization of rram and ots selector for advanced low voltage cmos compatibility," in *2020 IEEE International Memory Workshop (IMW)*, pp. 1–4, 2020.

[11] L. Goux, "Oxram technology development and performances," in *Advances in Non-Volatile Memory and Storage Technology (Second Edition)* (B. Magyari-Köpe and Y. Nishi, eds.), Woodhead Publishing Series in Electronic and Optical Materials, pp. 3–33, Woodhead Publishing, second edition ed., 2019.

[12] F. Lentz, B. Roesgen, V. Rana, D. J. Wouters, and R. Waser, "Current compliance-dependent nonlinearity in tio2 reram," *IEEE Electron Device Letters*, vol. 34, no. 8, pp. 996–998, 2013.

[13] G. Bersuker, D. Gilmer, and D. Veksler, "2 - metal-oxide resistive random access memory (rram) technology: Material and operation details and ramifications," in *Advances in Non-Volatile Memory and Storage Technology (Second Edition)* (B. Magyari-Köpe and Y. Nishi, eds.), Woodhead Publishing Series in Electronic and Optical Materials, pp. 35–102, Woodhead Publishing, second edition ed., 2019.

[14] A. Levisse, B. Giraud, J.-P. Noël, M. Moreau, and J. M. Portal, "High density emerging resistive memories: What are the limits?," *2017 IEEE 8th Latin American Symposium on Circuits & Systems (LASCAS)*, pp. 1–4, 2017.

[15] S. K. Kingra, V. Parmar, S. Negi, A. Bricalli, G. Piccolboni, A. Regev, J.-F. Nodin, G. Molas, and M. Suri, "Dual-configuration in-memory computing bitcells using SiOx RRAM for binary neural networks," *Applied Physics Letters*, vol. 120, 01 2022. 034102.

[16] C.-X. Xue, W.-H. Chen, J.-S. Liu, J.-F. Li, W.-Y. Lin, W.-E. Lin, J.-H. Wang, W.-C. Wei, T.-Y. Huang, T.-W. Chang, T.-C. Chang, H.-Y. Kao, Y.-C. Chiu, C.-Y. Lee, Y.-C. King, C.-J. Lin, R.-S. Liu, C.-C. Hsieh, K.-T. Tang, and M.-F. Chang, "Embedded 1-mb reram-based computing-in- memory macro with multibit input and weight for cnn-based ai edge processors," *IEEE Journal of Solid-State Circuits*, vol. 55, no. 1, pp. 203–215, 2020.

[17] J.-M. Hung, T.-H. Wen, Y.-H. Huang, S.-P. Huang, F.-C. Chang, C.-I. Su, W.-S. Khwa, C.-C. Lo, R.-S. Liu, C.-C. Hsieh, K.-T. Tang, Y.-D. Chih, T.-Y. J. Chang, and M.-F. Chang, "8-b precision 8-mb reram compute-in-memory macro using direct-current-free time-domain readout scheme for ai edge devices," *IEEE Journal of Solid-State Circuits*, vol. 58, no. 1, pp. 303–315, 2023.

[18] L. Grenouillet, N. Castellani, A. Persico, V. Meli, S. Martin, O. Billoint, R. Segaud, S. Bernasconi, C. Pellissier, C. Jahan, C. Charpin-Nicolle, P. Dezest, C. Carabasse, P. Besombes, S. Ricavy, N.-P. Tran, A. Magalhaes-Lucas, A. Roman, C. Boixaderas, T. Magis, M. Bedjaoui, M. Tessaire, A. Seignard, F. Mazen, S. Landis, E. Vianello, G. Molas, F. Gaillard, J. Arcamone, and E. Nowak, "16kbit 1t1r oxram arrays embedded in 28nm fdsoi technology demonstrating low ber, high endurance, and compatibility with core logic

transistors," in *2021 IEEE International Memory Workshop (IMW)*, pp. 1–4, 2021.

[19] W. Wan, R. Kubendran, C. Schaefer, S. B. Eryilmaz, W. Zhang, D. Wu, S. Deiss, P. Raina, H. Qian, B. Gao, S. Joshi, H. Wu, H.-S. P. Wong, and G. Cauwenberghs, "A compute-in-memory chip based on resistive random-access memory," *Nature*, vol. 608, no. 7923.

[20] G. W. Burr, R. S. Shenoy, and H. Hwang, "Select device concepts for crossbar arrays," in *Resistive Switching*, ch. 22, pp. 623–660, John Wiley & Sons, Ltd, 2016.

[21] A. Chen, "Memory select devices," in *Emerging Nanoelectronic Devices*, ch. 12, pp. 227–245, John Wiley & Sons, Ltd, 2014.

[22] A. Chen, "A review of emerging non-volatile memory (NVM) technologies and applications," *Solid State Electronics*, vol. 125, pp. 25–38, Nov. 2016.

[23] Y. S. Chen, H. Y. Lee, P. S. Chen, P. Y. Gu, C. W. Chen, W. P. Lin, W. H. Liu, Y. Y. Hsu, S. S. Sheu, P. C. Chiang, W. S. Chen, F. T. Chen, C. H. Lien, and M.-J. Tsai, "Highly scalable hafnium oxide memory with improvements of resistive distribution and read disturb immunity," in *2009 IEEE International Electron Devices Meeting (IEDM)*, pp. 1–4, 2009.

[24] B. Govoreanu, G. Kar, Y.-Y. Chen, V. Paraschiv, S. Kubicek, A. Fantini, I. Radu, L. Goux, S. Clima, R. Degraeve, N. Jossart, O. Richard, T. Vandeweyer, K. Seo, P. Hendrickx, G. Pourtois, H. Bender, L. Altimime, D. Wouters, J. Kittl, and M. Jurczak, "10×10nm2 hf/hfox crossbar resistive ram with excellent performance, reliability and low-energy operation," in *2011 International Electron Devices Meeting*, pp. 31.6.1–31.6.4, 2011.

[25] D. Ielmini, "Resistive switching memories based on metal oxides: mechanisms, reliability and scaling," *Semiconductor Science Technology*, vol. 31, p. 063002, June 2016.

[26] H.-S. P. Wong, H.-Y. Lee, S. Yu, Y.-S. Chen, Y. Wu, P.-S. Chen, B. Lee, F. T. Chen, and M.-J. Tsai, "Metal–oxide rram," *Proceedings of the IEEE*, vol. 100, no. 6, pp. 1951–1970, 2012.

[27] A. Prakash and H. Hwang, "Multilevel cell storage and resistance variability in resistive random access memory," *Physical Sciences Reviews*, vol. 1, no. 6, pp. –, 2016.

[28] J. Reuben, D. Fey, and C. Wenger, "A modeling methodology for resistive ram based on stanford-pku model with extended multilevel capability," *IEEE Transactions on Nanotechnology*, vol. 18, pp. 647–656, 2019.

[29] A. Prakash, J. Park, J. Song, J. Woo, E. Cha, and H. Hwang, "Demonstration of low power 3-bit multilevel cell characteristics in a taox-based rram by stack engineering," *IEEE Electron Device Letters*, vol. 36, pp. 32–34, Jan 2015.

[30] U. Russo, D. Kamalanathan, D. Ielmini, A. L. Lacaita, and M. N. Kozicki, "Study of multilevel programming in programmable metallization cell (pmc) memory," *IEEE Transactions on Electron Devices*, vol. 56, pp. 1040–1047, May 2009.

[31] S. Yu, Y. Wu, and H.-S. P. Wong, "Investigating the switching dynamics and multilevel capability of bipolar metal oxide resistive switching memory," *Applied Physics Letters*, vol. 98, p. 103514, 03 2011.

[32] A. Nitayama and H. Aochi, “Bit cost scalable (bics) technology for future ultra high density memories,” in *2013 International Symposium on VLSI Technology, Systems and Application (VLSI-TSA)*, pp. 1–2, 2013.

[33] Y. Deng, H.-Y. Chen, B. Gao, S. Yu, S.-C. Wu, L. Zhao, B. Chen, Z. Jiang, X. Liu, T.-H. Hou, Y. Nishi, J. Kang, and H.-S. P. Wong, “Design and optimization methodology for 3d rram arrays,” in *2013 IEEE International Electron Devices Meeting*, pp. 25.7.1–25.7.4, 2013.

[34] B. Hudec, C.-C. Chang, I.-T. Wang, K. Fröhlich, and T.-H. Hou, “Three dimensional integration of rerams,” in *2018 IEEE 18th International Conference on Nanotechnology (IEEE-NANO)*, pp. 1–2, 2018.

[35] K. Son, K. Cho, S. Kim, G. Park, K. Song, and J. Kim, “Modeling and signal integrity analysis of 3d xpoint memory cells and interconnections with memory size variations during read operation,” in *2018 IEEE Symposium on Electromagnetic Compatibility, Signal Integrity and Power Integrity*, pp. 223–227, 2018.

[36] D. Waddington, M. Kunitomi, C. Dickey, S. Rao, A. Abboud, and J. Tran, “Evaluation of intel 3d-xpoint nvdimm technology for memory-intensive genomic workloads,” in *Proceedings of the International Symposium on Memory Systems*, MEMSYS ’19, (New York, NY, USA), p. 277–287, Association for Computing Machinery, 2019.

[37] L. Zhang, S. Cosemans, D. J. Wouters, B. Govoreanu, G. Groeseneken, and M. Jurczak, “Analysis of vertical cross-point resistive memory (vrram) for 3d rram design,” in *2013 5th IEEE International Memory Workshop*, pp. 155–158, 2013.

[38] D. J. Wouters, R. Waser, and M. Wuttig, “Phase-change and redox-based resistive switching memories,” *Proceedings of the IEEE*, vol. 103, no. 8, pp. 1274–1288, 2015.

[39] B. Hudec, C.-W. Hsu, I.-T. Wang, W.-L. Lai, C.-C. Chang, T.-C. Wang, K. Frohlich, C. H. Ho, C.-H. Lin, and T. Hou, “3d resistive ram cell design for high-density storage class memory—a review,” *Science China Information Sciences*, vol. 59, pp. 1–21, 2016.

Part II
PERIPHERAL CIRCUITRY

Chapter 5
Circuits to Read from the ReRAM Array: Sense Amplifiers

5.1 Introduction

The peripheral circuitry consists of circuits around the memory array which are usually constructed using CMOS technology. By 'peripheral circuitry', we mean the circuits used to write data (*i.e.* to program the memory cells), to read out data (*i.e.* sense amplifiers used to convert the state of the memory cell to a voltage for further processing), the circuits used to select a row and a column (decoders/multiplexers) and any other additional circuits which are crucial while accessing the memory array (*e.g.* error detection/correction circuits). In short, the peripheral circuits form a channel through which the data in electronic form (voltage) is moved into the memory array (to be stored as resistance) and moved out of the memory array (from resistance to a CMOS-compatible voltage for further processing).

5.2 Read-out Circuits: Basic Principle of Sensing

As stated, the read-out circuits mainly serve the purpose of converting the state of the memory cell to a voltage which can be processed further. Traditionally, processing of data has been performed outside the memory*

* In this book, the memory is used only as a storage medium so far. The in-memory computing paradigm seeks to compute/process data at the location of the data *i.e.* the memory array itself. This is an important development in this field and this is discussed in Part III. Even if data is processed in the memory array,

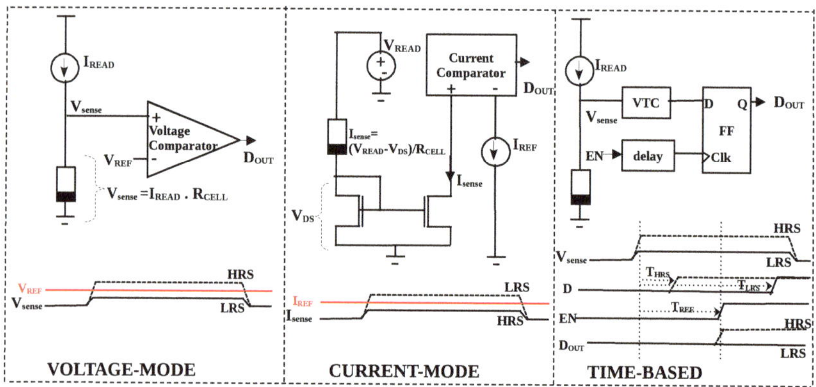

Fig. 5.1 Three modes of sensing the state of a ReRAM cell: In Voltage-mode sensing, V_{sense} is compared with V_{REF} to decipher the state of the cell ($V_{sense}(HRS) > V_{REF} > V_{sense}(LRS)$). In current-mode sensing, I_{sense} is compared with I_{REF} to determine the state of the cell ($I_{sense}(HRS) < I_{REF} < I_{sense}(LRS)$). In time-mode sensing, V_{sense} is converted into a proportional time delay by a Voltage-to-time Converter (VTC) and discriminated in time domain.

and the read-out circuit converts the data (stored as the state of the NVM) to a form amenable for CMOS processing, *i.e.* a voltage. In ReRAM technology, this means the HRS/LRS of the cell needs to be converted to a CMOS compatible voltage, *i.e.* V_{DD} or 0 V, corresponding to logic '1' and '0'. This can be performed in three ways – voltage-mode sensing, current-mode sensing and time-based sensing. Fig. 5.1 is a simplified illustration of these three ways. In voltage-mode sensing, the state of the cell is converted to a voltage by injecting a current I_{READ} through the ReRAM cell. This results in a voltage proportional to the resistance of the memory cell, $V_{sense}= I_{READ} \times R_{cell}$, where R_{cell} is either high or low resistance. V_{sense} is compared with a reference voltage V_{REF} using a voltage comparator to determine the state of the memory cell. A CMOS differential amplifier[†] can be used as a voltage comparator and its output changes to V_{DD} if $V_{sense} > V_{REF}$ or 0 V, if it is vice versa. In current-mode sensing, a small voltage

read-out circuits are needed to read them out. Furthermore, there are limitations to in-memory computing and all computing cannot be performed efficiently in the memory array and certainly a combination of memory and dedicated CMOS circuits are needed.

[†] A differential amplifier amplifies the difference between the voltage at its inputs and pulls the output to V_{DD} if the voltage at its positive terminal is greater than the voltage at its negative terminal, and 0 V if it is vice versa.

V_{READ} is applied across the cell. Since the ReRAM cell can switch at voltages near $V_{SET/RESET}$, care must be taken to make sure that V_{READ} is in safe limits below $V_{SET/RESET}$. The V_{READ} applied across the cell results in a current proportional to the resistance of the cell, $I_{sense}=\frac{V_{READ}-V_{DS}}{R_{cell}}$. The current I_{sense} is mirrored by a current mirror and fed into a current comparator, as depicted in Fig. 5.1. The current comparator outputs V_{DD} if $I_{sense} > I_{REF}$ and 0 V, if it is vice versa.

The time-based sensing scheme is very similar to voltage-mode sensing since the resistance state is converted to an equivalent voltage V_{sense} as it was the case in voltage-mode sensing. The difference lies in the way the V_{sense} is processed. Instead of comparing with V_{REF}, the V_{sense} is converted to a time delay using a Voltage-to-Time Converter (VTC). As depicted in Fig. 5.1, the VTC converts V_{sense} to T_{HRS} and T_{LRS}. A D flip-flop which has an EN signal going high at T_{REF} ($T_{HRS} < T_{REF} < T_{LRS}$) differentiates between the two states. Unlike the voltage-mode sensing, in time-mode sensing, the difference in V_{sense} (for HRS and LRS) is converted to time-delay and differentiated in time domain using a D flip-flop.

In Fig. 5.1, the read-out circuits are shown connected to an individual memory cell. In practice, the read-out circuits have to be connected to the memory array (*i.e.* BL or SL) and the simplified circuits of Fig. 5.1 cannot be used directly. Significant modifications are needed and in the following sections, we will discuss how each of these sensing types can be adapted to sense a single memory cell which is located in a larger memory array. We will also elaborate the detailed CMOS implementation of the blocks shown in Fig. 5.1. In this chapter, we will discuss read-out circuits for 1T-1R arrays and the same circuits can be used to sense data from 1S–1R arrays with little or no modifications.

5.3 Voltage-mode Sensing

To sense the state of the cell using the voltage-mode SA presented in Fig. 5.1, I_{READ} must be injected for a considerable time t_{sense} till V_{sense} develops to a substantial voltage, *i.e.* there must be enough difference between V_{sense} for HRS and LRS so that they can be differentiated by the voltage comparator ($V_{sense}(\text{HRS}) > V_{REF} > V_{sense}(\text{LRS})$). This will consume considerable energy, $I_{READ} \int_0^{t_{sense}} V_{sense}(t).dt$. In other words, there will be static power dissipation throughout the duration of sensing operation. To overcome this, it is customary to pre-charge a node and make that node discharge through the ReRAM cell for a small duration. The pre-charged node will discharge according to the state of the cell and consequently, the voltage across the cell will drop according to the state of the cell, which can be sensed.

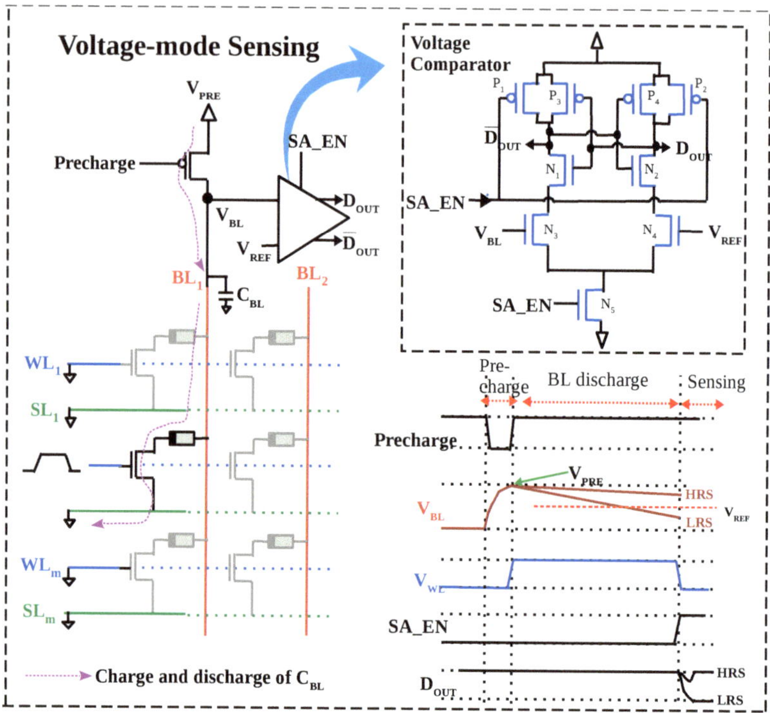

Fig. 5.2 The BL is first pre-charged to a specific voltage, V_{PRE}. The pre-charged BL discharges according to the resistance of the memory cell during BL discharge phase. When SA_EN goes high, $\overline{D_{OUT}}$ discharges faster if V_{BL} is greater than V_{REF}. The faster discharge of the voltage at $\overline{D_{OUT}}$ is reinforced by the positive feedback formed by the cross-coupled inverters. Hence, D_{OUT} remains high and $\overline{D_{OUT}}$ goes low for HRS sensing.

Fig. 5.2 depicts the detailed circuit to read from a 1T–1R array using voltage-mode sensing technique and the conceptual wave-forms. The PMOS transistor above the BL_1 is used to pre-charge BL_1 to a specific voltage V_{PRE}. It must be noted that the BL is a long wire running through the entire memory array. Depending on the size of the memory array and the metal-pitch, the BL parasitic capacitance can vary between few fF to pFs. This is denoted by C_{BL} and is charged to V_{PRE} during Pre-charge phase. It must be noted that all the SLs and WLs of the array are connected to ground during READ operation, as shown in Fig. 5.2. Now, the WL of the cell which has to be sensed, is activated. At this juncture, C_{BL} which was pre-charged to V_{PRE} starts to discharge in accordance with the resistance/state of the cell. Notice that although BL_1 is connected to all the memory cells in column 1, there is no other path for C_{BL} to discharge since WL_{1-m} are grounded (access transistors of all other memory cells in column 1 are OFF). If the cell is in HRS, the discharge rate is slow and $V_{BL} \approx V_{PRE}$. If the cell is in LRS, the rate of discharge is faster and V_{BL} is much below V_{PRE}. At the end of BL discharge phase, the WL is deactivated and hence no more discharge is possible. The V_{BL} is now stable and can be sensed by comparing it with a reference voltage, V_{REF}. A CMOS voltage comparator can be used to compare V_{BL} and V_{REF}. The comparator depicted was proposed by T. Kobayashi et al. of Toshiba Corporation in 1993 [1]. Popularly called the Strong ARM latch[‡], the circuit can sense very minute difference in voltages very efficiently while consuming zero static power [2].

The sensing phase starts when SA_EN goes high. Before SA_EN goes high, nodes D_{OUT} and $\overline{D_{OUT}}$ are pre-charged to V_{DD} through P_1 and P_2. Also, V_{BL} and V_{REF} are connected to two NMOS transistors N_3 and N_4, whose source terminals are connected together. Therefore, the currents through N_3 and N_4 will be a function of their gate voltages only. When SA_EN goes high, transistors N_3 and N_4 conduct current since there is a path to conduct current through N_5. If $V_{BL} > V_{REF}$, N_3 conducts more current than N_4. Therefore node $\overline{D_{OUT}}$ begins to discharge faster than node D_{OUT}. However, $\overline{D_{OUT}}$ is the input for the inverter formed by P4–N2. Hence the drop in voltage at $\overline{D_{OUT}}$ leads to a increase in voltage at node D_{out}. This, in turn, reinforces the low at $\overline{D_{OUT}}$ because D_{out} is the input to the inverter formed by the P3–N1 pair. In other words, when SA_EN goes high, the faster discharge of the voltage at $\overline{D_{OUT}}$ is reinforced by the positive feedback formed by the cross-coupled inverters. Hence, D_{OUT} remains high and $\overline{D_{OUT}}$ goes low. In this manner, for HRS, V_{BL} being greater than V_{REF} is sensed as a logic HIGH at D_{OUT}. Following a similar analysis, LRS case which results in V_{REF} being greater than V_{BL} will be sensed as a logic LOW at D_{OUT}.

‡ This comparator became known to be so due to its use in the StrongARM microprocessor of DEC

In practice, the voltage-mode SA of Fig. 5.2 will need to be custom designed to meet the requirements of a particular ReRAM technology and size of the memory, *i.e.* C_{BL} will decide how long the pre-charge pulse should be and C_{BL} depends on the size of the memory array and the parasitic capacitance of the metal lines used for BL. Similarly, HRS/LRS will decide how long the WL should be activated since discharge time is a function of $R_{cell} \cdot C_{BL}$. Furthermore, read-disturb (discussed in Section 5.6.1) phenomenon necessitates that V_{PRE} be < 0.3 V so that the state of the cell is not altered during READ operation. Let us assume that V_{PRE} is 300 mV and after BL discharge, BL drops to 280 mV and 220 mV for HRS and LRS, respectively. Then V_{REF} can be chosen to be 250 mV so that there is a 30 mV margin between V_{BL} and V_{REF} in both cases. Depending on the CMOS process node in which the SA is designed, such low voltages may not be enough to turn on N_3, N_4 for sensing and voltage comparator may need to be modified, *e.g.* N_3, N_4 can be made wide to conduct reasonable current in sub-threshold region. In summary, the voltage-mode SA needs to be re-designed for a particular ReRAM cell's HRS, LRS, BL parasitic capacitance and is not a one-design fits-all solution for sensing. Further improvements to this basic voltage-mode sensing technique can be found in [3, 4]. A design example in CMOS 130 nm process technology can be found at the end of this chapter.

5.4 Current-mode Sensing

In current-mode sensing, the voltage across the memory cell should be kept constant for some duration so that the memory cell's resistance can be converted to an equivalent current. The resulting cell current can then be detected by a current sense amplifier. Fig. 5.3 depicts a practical current-mode SA for ReRAM technology. Current-mode sensing, as stated before, is performed by reading out the current from the memory cell and comparing it with a reference current, I_{REF} using a current comparator. This necessitates a constant voltage V_{READ} to be applied across the cell during sensing. If V_{READ} is applied at the BL, then the SL must be grounded so that V_{READ} appears across the 1T–1R cell. But the current sense amplifier cannot be connected to the SL/ground since it needs to sink in current and must be biased at a higher voltage. Therefore, one approach is to bias the SL at V_{BIAS} using an Error Amplifier (EA_1), as depicted in Fig. 5.3. Biasing of SL_1 at V_{BIAS} using EA_1 can be understood as follows.

If the gain of Error Amplifier EA_1 is very high, even a small change in the differential voltage input to the amplifier introduces a significant change in the output voltage. Because we have a negative feedback loop[§], the loop does not allow such a big change. That means, the loop always keeps the differential voltage small enough so that the voltage at the drain of N_5 (or the non-inverting input of error amp) is very close to V_{BIAS}. The error amplifiers in Fig. 5.3 can be implemented using a basic differential amplifier.

Since $V_{READ}+V_{BIAS}$ is applied at BL_1, V_{READ} appears across the 1T-1R cell resulting in a current, $I_{sense} = \frac{V_{READ}}{R_{cell}}$ (drain of N_5 is biased at V_{BIAS}). I_{sense} flows into transistor N_5. Transistors $N_5 - N_6$ form a current mirror and therefore N_6 sinks I_{sense}. Similarly, another error amplifier EA_2 biases the drain of N_8 at V_{BIAS}. Therefore, $I_{REF} = \frac{V_{DD}-V_{BIAS}}{R_{REF}}$ and a pre-determined I_{REF} can be set precisely using R_{REF}. Finally, I_{sense} and I_{REF} are compared using the current comparator. The current comparator functions similar to the voltage comparator of section 5.3. Initially SA_{EN} is low and nodes D_{OUT} and $\overline{D_{OUT}}$ are pre-charged to V_{DD}. When SA_{EN} goes high, transistors N_3 and N_4 are ON creating a path for nodes D_{OUT} and $\overline{D_{OUT}}$ to discharge. One of them discharges at a faster rate, which is reinforced by the positive feedback formed by cross-coupled inverters [5, 6]. For example, if I_{sense} is greater than I_{REF}, $\overline{D_{OUT}}$ discharges faster than D_{OUT}. As $\overline{D_{OUT}}$ tends to a lower voltage, it tends to pull D_{OUT} node to a higher voltage due to the inverter formed by $P_4 - N_2$. This tendency to pull D_{OUT} high magnifies the tendency to push node $\overline{D_{OUT}}$ to logic LOW through the $P_3 - N_1$ inverter. In a very short time, D_{OUT} is pulled to V_{DD} and $\overline{D_{OUT}}$ to 0 V. It must be noted that, at the start of the sensing phase, both D_{OUT} and $\overline{D_{OUT}}$ tend to logic LOW transiently before the positive feedback kicks in, as depicted in the conceptual wave-forms. However, in a few ns, the stronger discharge path dominates in pulling the node to ground. Thus, $I_{sense} > I_{REF}$ is sensed as a logic HIGH and $I_{sense} < I_{REF}$ is sensed as a logic LOW at D_{OUT} node.

[§] Negative feedback in electronic circuits is a self-correcting mechanism in which the circuit behaves in such a way as to oppose a change. Suppose the drain voltage of N_5 **increases** by 2%. Since the non-inverting input of EA_1 has increased by 2%, the output of the error amplifier increases according to its gain, $V_{OUT,EA1} = A\times(V^+ - V^-)$. Now the current through N_5 increases since its gate voltage is the error-amplifier's output. An increased current through N_5 increases the voltage drop across R_{CELL} and **decreases** V_{DS} of N_5 (Voltage drop across $R_{CELL}+V_{DS,N5}$ must be equal to $V_{READ}+V_{BIAS}$). In this manner, the initial change (increase in drain voltage of N_5) is corrected and the circuit is stabilized.

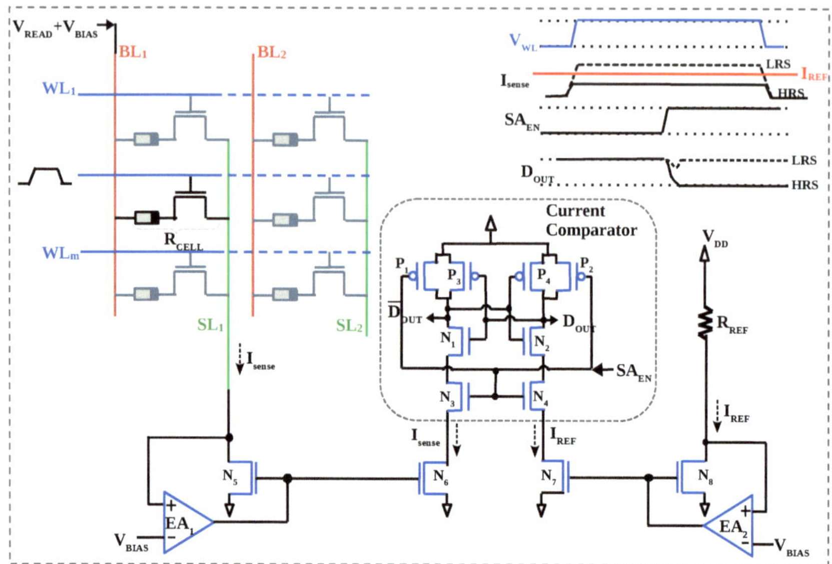

Fig. 5.3 In current-mode sensing, a constant voltage is applied across the cell and the current I_{sense} from the memory cell is mirrored and then compared with I_{REF} to determine the state of the cell ($I_{sense}(HRS) < I_{REF} < I_{sense}(LRS)$).

The current-mode SA consumes more peripheral circuit area compared to the voltage-mode SA of Fig. 5.2 due to two error amplifiers needed to bias the SL at a fixed voltage (also called clamping because the SL is clamped to a particular voltage). But the circuit is very effective in sensing even tens of nA difference between I_{sense} and I_{REF}. Regarding power consumption, it must be noted that current-mode SA draws static power supply through the sense and reference currents. It must be noted that although I_{sense} is shown flowing through N_5 and N_6 in Fig. 5.3, I_{sense} flows only through N_5 throughout the sensing process *i.e.* as long as WL remains high. The mirrored I_{sense} flows through N_6 only for ns duration when SA_{EN} goes high (when SA_{EN} is low, N_3, N_4 are OFF and cannot conduct I_{sense} and I_{REF}). This is because the current comparator has inherent cut-off capability once one of the nodes is pulled to ground. For example, if $\overline{D_{OUT}}$ is pulled to ground, D_{OUT} will be high and can still sink in current as long as SA_{EN} is high. But $\overline{D_{OUT}}$ being ground implies that N_2 is OFF cutting off a discharge path for I_{REF} to flow. In this manner, the current comparator is very energy efficient and does not conduct static currents throughout the time SA_{EN} is high. However, for the duration for which WL is high, there will be static currents I_{sense} and I_{REF} through N_5 and N_8, respectively.

Notice that we have followed SL clamping to read-out a constant cell current. In literature, many current mode SAs are reported which perform BL clamping instead of SL clamping [7] *i.e.* the BL is clamped at a fixed voltage instead of the SL. The interested reader is referred to [8, 4, 5, 6, 7] for different implementations of current-mode sense amplifiers.

5.5 Time-based Sensing

Both the sensing techniques described so far need an absolute reference voltage/current to perform comparison with the sensed voltage/current. Generating precise reference (V_{REF}/I_{REF}) is a non-trivial task and also adds overhead to the sensing circuitry. Inspired by time-domain computing [9], researchers have proposed sensing circuits based in time-domain. These circuit are called time-based sensing circuits and do not require an absolute voltage or current reference. These circuits convert the state of the ReRAM cell into a proportional voltage/current. But instead of comparing with a reference voltage/current, they convert the quantity to be sensed into time and compare in time-domain. Voltage-to-time (VTC) or Current-to-time (CTC) converters are employed to convert the voltage/current and these can be implemented as simple CMOS circuits. Time-based sensing does need a time-reference for which the clock signal of the memory system can be used with appropriate processing (usual the clock signal is delayed and used as a reference). The greatest advantage of time-based sensing is that they do not require the generation of voltage/current reference circuits on-chip, thus saving area and power. However, time-based sensing circuit are more prone to PVT variations and must be carefully designed. In this section, we will discuss a basic time-based sensing circuit which converts the resistance of the cell into a BL voltage which is then converted to a time delay and sensed in time-domain. This sensing scheme was originally proposed to read data from STT-MRAM [10], which have a resistance window of a few KΩ.

The time-based sensing circuit is essentially a voltage-to-time converter followed by a time-domain comparator (D-flip flop). Voltage-to-time conversion is achieved by the current-starved inverter (transistors M_{1-5}) followed by transistor M_6 and an inverter (Fig. 5.4). Similar to

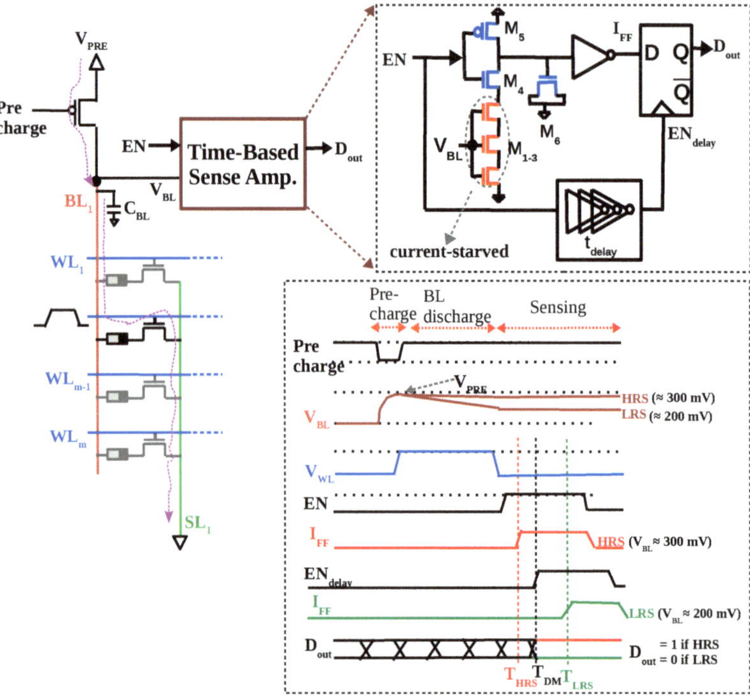

Fig. 5.4 Similar to voltage-mode sensing, the BL is pre-charged and discharged in time-based sensing to convert the resistance to a voltage V_{BL}. A current-starved inverter transforms this voltage into a proportional delay which is sensed as a CMOS-compatible voltage by the D-FF [10] .

the voltage-mode sensing, the BL is pre-charged to a particular voltage (*e.g.* V_{PRE}= 300 mV). Then, the WL of the cell which is to be sensed is activated (goes HIGH). This forms a path for the BL to discharge. As depicted in Fig. 5.4, the capacitance of BL_1 begins to discharge through the 1T-1R cell to ground. If the cell is in HRS, the BL discharges very slowly and stays ≈ 300 mV. On the other hand, if the cell is in LRS, the BL discharges rapidly. The WL is activated for a particular duration (BL discharge phase) so that there is a discernible difference in V_{BL} for the two cases at the end of the BL discharge phase. At the end of BL discharge phase, WL is de-activated and V_{BL} is stable (approximately 300 mV if the cell was in HRS and 200 mV if the cell was in LRS) and ready to be sensed.

The sensing phase starts when EN goes high. As stated, transistors $M_5 - M_4$ behave as an inverter and start to transmit EN signal. But the gate of transistors $M_{1,2,3}$ are connected to V_{BL} and operate in sub threshold region since V_{BL} is 200 mV/300 mV. Such a V_{BL} (few hundred mV) limits the current flow through the inverter (transistor M_{1-5}), hence the name current-starved inverter. When EN goes high, the current-starved inverter introduces a delay proportional to V_{BL}, *i.e.* a higher V_{BL} incurs less delay. A V_{BL} of 300 mV incurs less delay and low-to-high transition of EN reaches the input of the flip-flop (I_{FF}) faster, *i.e.* at T_{HRS}. For a lower V_{BL} of 200 mV, the delay is greater and the low-to-high transition occurs at T_{LRS}. For best results, V_{BL} must be approximately equal to V_{TH} (threshold voltage of NMOS) so that it conducts faster for HRS case and V_{BL} must be below V_{TH} so that it conducts slower for LRS case. Since the current-starved transistors M_{1-3} are the crucial factor in deciding the delay, they can be made large to make the circuit less sensitive to CMOS process variations. EN_{delay}, the EN signal delayed by t_{delay} acts as the edge trigger for the D-FF. When EN_{delay} goes high at T_{DM} (Decision Moment), it latches the signal at I_{FF} and hence the flip flop output D_{out} is high for HRS and low for LRS. t_{delay} can be set precisely in between T_{HRS} and T_{LRS} by introducing a delay in EN signal. Thus, EN_{delay} which acts as a reference for time-based sensing can be generated from the EN signal itself¶. More variations and improvements to time-based sensing are presented in [11, 12, 13, 14].

5.6 ReRAM technology-specific issues to be considered while sensing

5.6.1 Read-disturb

Read-disturb refers to the perturbation of the conductive filament during the data reading phase resulting in accidental change of stored memory state [15, 16]. Experimental work in [17] suggests that read-disturb occurs only when the READ voltage across the ReRAM cell is larger than 0.3 V. A similar experimental work by a different group [7] also confirmed that the read-disturb phenomenon is severe at READ voltage above 0.3 V. Hence,

¶ Time delay can be generated in CMOS by many methods, the simplest of which is just a series of inverters with MOS capacitive loads between them

while reading, one must make sure that the voltage across the 1T-1R cell is below 0.3 V. In voltage-mode and time-based sensing discussed in this chapter, the reader might recall that the BL was pre-charged to a particular voltage V_{PRE} and then dis-charged through the ReRAM cell. To avoid read-disturb, V_{PRE} must be 0.3 V or less. This is because V_{PRE} will appear across the 1T-1R cell at the beginning of the discharge phase. The reader might recall that, to sense a cell, the WL is activated so that the precharged BL can discharge to ground through the 1T-1R cell (SL is at ground during discharge phase, Fig. 5.2, Fig.5.4). Since V_{PRE} is the maximum voltage that can appear across the 1T-1R cell while sensing, limiting V_{PRE} to be ≤ 0.3 V will make sure that read-disturb does not occur while reading the cell. In current-mode sensing, the SL is held at V_{BIAS} and $V_{READ} + V_{BIAS}$ is applied to the BL (see Fig. 5.3). In this case, V_{READ} must not be larger than 0.3 V although the BL can be biased at higher voltages *e.g.* biasing the BL at 1.3 V and the SL at 1 V will result in safe reading since only 0.3 V appears across the 1T-1R cell.

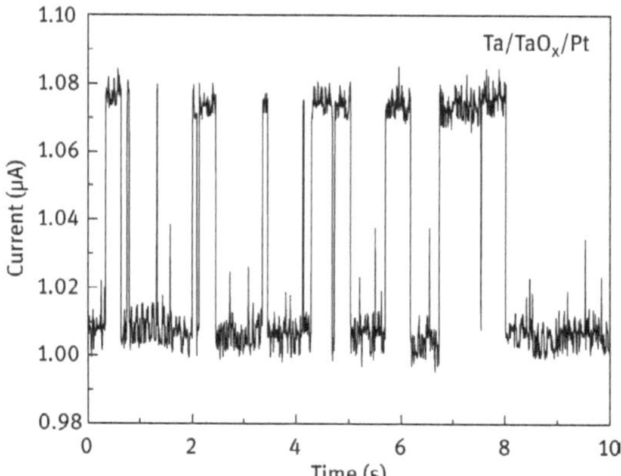

Fig. 5.5 Random Telegraph Noise (RTN observed in Ta/TaOx/Pt device. Reprinted from [18]

5.6.2 Random Telegraph Noise (RTN)

Strictly speaking, RTN is a phenomenon which is observed in many nano-scale devices including CMOS transistors in sub-14 nm nodes [19, 20]. In ReRAM, Random Telegraph Noise has been observed while reading the state of ReRAM cell [18]. Severe fluctuations were observed in the read current amplitude and this causes a reduction in the read margin. RTN is a low frequency noise affecting the efficiency of the read-out process since any noise affects the signal-to-noise ratio [20]. Typically, the read-out current switches between two stable values. Fig. 5.5 depicts the RTN observed in a Ta/TaOx/Pt device [18]. As can be observed, the read-out current at HRS switches between 1 μA and 1.08 μA. The origin of the RTN in ReRAM is attributed to the capture and emission (trap and detrap) of electrons in the trap (more probably oxygen vacancy trap) near the filament in LRS state and in the tunneling gap in the HRS state [21, 22].

5.7 Sense Amplifier Design Example

Problem: Design a **voltage-mode SA** for a ReRAM technology which has a HRS and LRS of 200 kΩ and 10 kΩ, respectively. The cells have a SET/RESET voltage of $\approx \pm 1$ V (The values are based on an ReRAM technology reported in [23]). Assume a BL capacitance of 0.25 pF. Design the sensing circuit in 130 nm CMOS.
Solution: Since the ReRAM device has a SET/RESET voltage of ± 1 V, it would be safe to use a maximum READ voltage of 0.3 V. Hence, V_{PRE} = 300 mV. Since the C_{BL} = 0.25 pF, the duration of the pre-charge pulse should be long enough to charge this bit-line capacitance. The NMOS transistor above BL_1 is used to pre-charge BL_1 (bit line parasitic capacitance C_{BL}) to 0.3 V. When the BL is precharged to V_{PRE} and discharged for a duration of $t_{discharge}$, the BL voltage would be

$$V_{BL} = V_{PRE} \cdot e^{\frac{-t_{discharge}}{R_{cell} \cdot C_{BL}}} \tag{5.1}$$

Depending on the resistance R_{cell} of the ReRAM cell, the BL gets discharged to different voltages for the same $t_{discharge}$. Let us assume a $t_{discharge}$ of 5 ns. For R_{cell}=200 kΩ,

$$V_{BL} = 300 \quad mV \cdot e^{\frac{-5 \cdot ns}{200k\Omega.0.25pF}} \approx 270 \quad mV \tag{5.2}$$

Similarly, for R_{cell}=10 kΩ, V_{BL} gets discharged to 40 mV. At the end of BL discharge phase, the WL is deactivated and hence no more discharge is possible. The V_{BL} is now stable and can be sensed by comparing it with a reference voltage, V_{REF} using a Sense Amplifier(SA). For C_{BL} of 0.25 pF, 5 ns was enough to charge the BL from 0 V to 300 mV. $t_{discharge}$ can be chosen to be 5 ns since 5 ns gives a margin ΔV_{BL} of 230 mV, which is good enough for sensing (270 mV- 40 mV). A V_{REF} of 155 mV can be used to distinguish between HRS and LRS. The BL was pre-charged for 5 ns and discharged for 5 ns, as depicted in Fig.5.6. The sensing phase was 5 ns making the total READ latency to be 15 ns. Since V_{BL} after discharge is very small (270 mV or 40 mV), it is not enough to turn ON the NMOS transistor for sensing if the circuit in Fig.5.2 is used. Therefore, we used a PMOS version of the voltage-mode SA presented in 5.2. As can be observed in Fig.5.6, V_{BL} is fed to PMOS transistor P_1 which forms the sense path and V_{REF} is fed to another PMOS transistor P_2 which forms the reference path. D_{OUT} and $\overline{D_{OUT}}$ are both at ground initially because SA_{EN} is HIGH. During sensing phase, SA_{EN} goes LOW and both the sense and reference paths are ON. If $V_{BL} < V_{REF}$, $\overline{D_{OUT}}$ charges faster than D_{OUT} and D_{OUT} is pulled to ground. In this manner, LRS is sensed as LOW at D_{OUT}. By a similar analysis, HRS is sensed as HIGH at D_{OUT}. The schematic and the conceptual waveforms are plotted in Fig. 5.6.

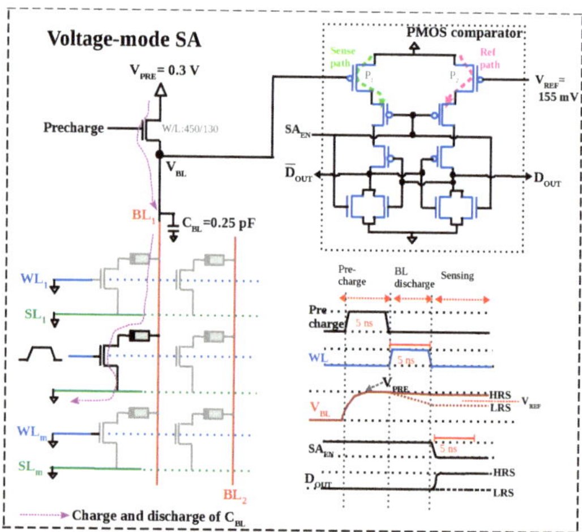

Fig. 5.6 Schematic of the voltage-mode SA simulated in CMOS 130 nm process of IHP. Except the pre-charge transistor, all transistors have a W/L of $\frac{300}{130}$.

References

[1] T. Kobayashi, K. Nogami, T. Shirotori, and Y. Fujimoto, "A current-controlled latch sense amplifier and a static power-saving input buffer for low-power architecture," *IEEE Journal of Solid-State Circuits*, vol. 28, no. 4, pp. 523–527, 1993.

[2] B. Razavi, "The strongarm latch [a circuit for all seasons]," *IEEE Solid-State Circuits Magazine*, vol. 7, no. 2, pp. 12–17, 2015.

[3] A. Lee, H. Lee, F. Ebrahimi, B. Lam, W.-H. Chen, M.-F. Chang, P. K. Amiri, and K.-L. Wang, "A dual-data line read scheme for high-speed low-energy resistive nonvolatile memories," *IEEE Transactions on Very Large Scale Integration (VLSI) Systems*, vol. 26, no. 2, pp. 272–279, 2018.

[4] M.-F. Chang, A. Lee, P.-C. Chen, C. J. Lin, Y.-C. King, S.-S. Sheu, and T.-K. Ku, "Challenges and circuit techniques for energy-efficient on-chip nonvolatile memory using memristive devices," *IEEE Journal on Emerging and Selected Topics in Circuits and Systems*, vol. 5, no. 2, pp. 183–193, 2015.

[5] W. Bae, K. J. Yoon, C. S. Hwang, and D.-K. Jeong, "A crossbar resistance switching memory readout scheme with sneak current cancellation based on a two-port current-mode sensing," *Nanotechnology*, vol. 27, p. 485201, oct 2016.

[6] W. Bae, K. J. Yoon, T. Song, and B. Nikolić, "A variation-tolerant, sneak-current-compensated readout scheme for cross-point memory based on two-port sensing technique," *IEEE Transactions on Circuits and Systems II: Express Briefs*, vol. 65, pp. 1839–1843, Dec 2018.

[7] M.-F. Chang, S.-S. Sheu, K.-F. Lin, C.-W. Wu, C.-C. Kuo, P.-F. Chiu, Y.-S. Yang, Y.-S. Chen, H.-Y. Lee, C.-H. Lien, F. T. Chen, K.-L. Su, T.-K. Ku, M.-J. Kao, and M.-J. Tsai, "A high-speed 7.2-ns read-write random access 4-mb embedded resistive ram (reram) macro using process-variation-tolerant current-mode read schemes," *IEEE Journal of Solid-State Circuits*, vol. 48, no. 3, pp. 878–891, 2013.

[8] J. Yin, C. Dou, D. Dong, J. Yu, X. Xu, Q. Luo, T. Gong, L. Tai, P. Yuan, X. Xue, M. Liu, and H. Lv, "A 0.75 v reference clamping sense amplifier for low-power high-density reram with dynamic pre-charge technique," *IEICE Electronics Express*, vol. 16, no. 12, pp. 20190201–20190201, 2019.

[9] Y. Zhang, J. Wang, C. Lian, Y. Bai, G. Wang, Z. Zhang, Z. Zheng, L. Chen, K. Zhang, G. Sirakoulis, and Y. Zhang, "Time-domain computing in memory using spintronics for energy-efficient convolutional neural network," *IEEE Transactions on Circuits and Systems I: Regular Papers*, vol. 68, no. 3, pp. 1193–1205, 2021.

[10] Q. Trinh, S. Ruocco, and M. Alioto, "Time-based sensing for reference-less and robust read in stt-mram memories," *IEEE Transactions on Circuits and Systems I: Regular Papers*, vol. 65, pp. 3338–3348, Oct 2018.

[11] J. Reuben and S. Pechmann, "A parallel-friendly majority gate to accelerate in-memory computation," in *2020 IEEE 31st International Conference on Application-specific Systems, Architectures and Processors (ASAP)*, pp. 93–100, 2020.

[12] J. Reuben and D. Fey, "A time-based sensing scheme for multi-level cell (mlc) resistive ram," in *2019 IEEE Nordic Circuits and Systems Conference*

(NORCAS): NORCHIP and International Symposium of System-on-Chip (SoC), pp. 1–6, 2019.

[13] J. Reuben and S. Pechmann, "Accelerated addition in resistive ram array using parallel-friendly majority gates," *IEEE Transactions on Very Large Scale Integration (VLSI) Systems*, vol. 29, no. 6, pp. 1108–1121, 2021.

[14] X. Zhang, B.-K. An, and T. T.-H. Kim, "A robust time-based multi-level sensing circuit for resistive memory," *IEEE Transactions on Circuits and Systems I: Regular Papers*, vol. 70, no. 1, pp. 340–352, 2023.

[15] N. Raghavan, "Statistics of disturb events in oxram devices — a phenomenological model," in *2017 IEEE International Reliability Physics Symposium (IRPS)*, pp. PM–3.1–PM–3.7, 2017.

[16] M. Zhao, B. Gao, J. Tang, H. Qian, and H. Wu, "Reliability of analog resistive switching memory for neuromorphic computing," *Applied Physics Reviews*, vol. 7, p. 011301, 01 2020.

[17] W. Shim, Y. Luo, J.-S. Seo, and S. Yu, "Investigation of read disturb and bipolar read scheme on multilevel rram-based deep learning inference engine," *IEEE Transactions on Electron Devices*, vol. 67, no. 6, pp. 2318–2323, 2020.

[18] A. Prakash and H. Hwang, "Multilevel cell storage and resistance variability in resistive random access memory," *Physical Sciences Reviews*, vol. 1, no. 6, pp. –, 2016.

[19] S. Dongaonkar, M. D. Giles, A. Kornfeld, B. Grossnickle, and J. Yoon, "Random telegraph noise (rtn) in 14nm logic technology: High volume data extraction and analysis," in *2016 IEEE Symposium on VLSI Technology*, pp. 1–2, 2016.

[20] J. Kim, H. Nili, N. D. Truong, T. Ahmed, J. Yang, D. S. Jeong, S. Sriram, D. C. Ranasinghe, S. Ippolito, H. Chun, and O. Kavehei, "Nano-intrinsic true random number generation: A device to data study," *IEEE Transactions on Circuits and Systems I: Regular Papers*, vol. 66, no. 7, pp. 2615–2626, 2019.

[21] S. Ambrogio, S. Balatti, A. Cubeta, A. Calderoni, N. Ramaswamy, and D. Ielmini, "Statistical fluctuations in hfox resistive-switching memory: Part ii—random telegraph noise," *IEEE Transactions on Electron Devices*, vol. 61, no. 8, pp. 2920–2927, 2014.

[22] F. M. Puglisi, P. Pavan, L. Larcher, and A. Padovani, "Analysis of rtn and cycling variability in hfo2 rram devices in lrs," in *2014 44th European Solid State Device Research Conference (ESSDERC)*, pp. 246–249, 2014.

[23] S. Pechmann, T. Mai, M. Völkel, M. K. Mahadevaiah, E. Perez, E. Perez-Bosch Quesada, M. Reichenbach, C. Wenger, and A. Hagelauer, "A versatile, voltage-pulse based read and programming circuit for multi-level rram cells," *Electronics*, vol. 10, no. 5, 2021.

Chapter 6
Circuits to Program the ReRAM Array: WRITE Circuits

6.1 Introduction

The reader might recall from Chapter 2 that the cell is written (*i.e.* its state changed from HRS to LRS or vice versa) by the application of appropriate voltage across the ReRAM cell's electrodes. The voltages V_{SET} and V_{RESET} are characteristics of the device and are determined by the switching oxide, its thickness, electrode material and other such physical properties of the ReRAM cell. *e.g.* a $Ti/SiO_2(5nm)/C$ device has a V_{SET} of 2.4 V and a V_{RESET} of -1.25 V, as reported in [1]. This means that, if the device is in HRS and 2.5 V (or any voltage greater than 2.4 V) is applied, the device will switch to LRS. Similarly, if the device is in LRS, the ReRAM cell is supposed to switch to HRS if any voltage of opposite polarity greater than 1.25 V is applied. However, this is an ideal situation. In practise, V_{SET} and V_{RESET} are not single values, but a range. Fig. 6.1 plots this reality, where the variation in V_{SET} and V_{RESET} is simulated by fitting the Stanford-PKU model to the variability observed in hafnium oxide ReRAMs manufactured at IHP Microelectronics, Germany [2].

This V_{SET} and V_{RESET} variation is a very important phenomenon to be kept in mind while writing to the cell because it affects the design of the WRITE circuitry. The variation has energy implications as well because, as we will see in the coming section, writing is performed by a series of pulses (instead of a single pulse) to minimize variation. To make matters simple, first we start with the basic WRITE circuitry assuming absence of variations. Therefore, the cell can be written by a single pulse. In fact, when ReRAM technology was first explored, the cells were programmed by a single pulse. Later, since programming by a single pulse

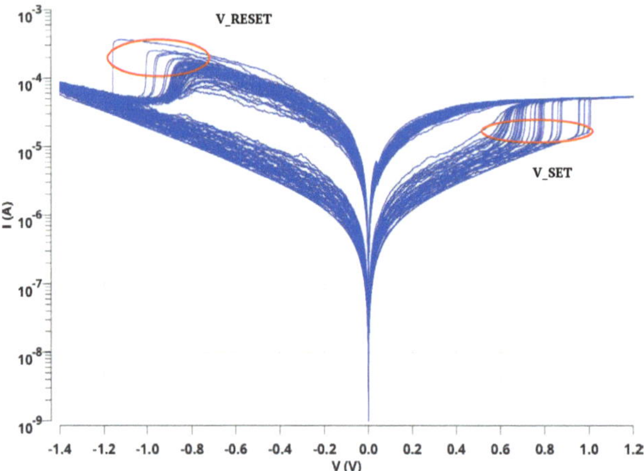

Fig. 6.1 In ReRAM technology, due to cycle-to-cycle variations, the device does not always switch from HRS to LRS and vice versa, at a particular voltage. The threshold voltage above which the device switches is not a single value, but a range. Redrawn from [2].

resulted in a lot of variations, two directions were pursued- incremental programming (series of pulses of progressively increasing amplitude with READ pulses in between) and self-termination programming (the WRITE stimulus is automatically cut-off once the cell is programmed to a target resistance).

6.2 Basic WRITE Circuit

We recall from Chapter 4 that when ReRAM cell is integrated in series with a transistor, it enables better programming since the current through the ReRAM cell is controlled by the transistor. This is especially crucial during SET process when the cell is switched from HRS to LRS. When the cell switches from HRS to LRS, the sudden lowering of resistance can cause a sudden increase in flow of current through the device – a current spike. The transistor avoids this current spike by limiting the current that can flow through the 1T-1R device to a few hundreds of μA, the maximum drain-source current corresponding to the gate voltage of the transistor. Due to the aforementioned reason, ReRAM manufactured in 1T-1R configuration is safer to operate compared to 1S-1R configuration.

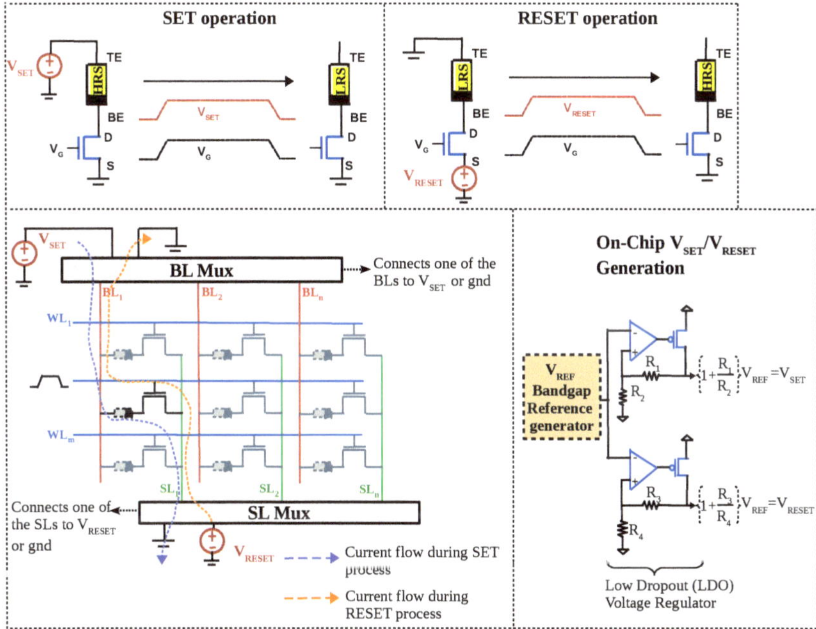

Fig. 6.2 SET and RESET process require voltage of opposite polarity across the device. The SET and RESET voltages can be generated on-chip from a band gap voltage reference using resistors and regulated by a Low dropout voltage regulator [3].

In this section, we discuss how a single 1T-1R cell is written or programmed to a specific state. Let us assume that the ReRAM cell is initially at HRS and needs to be programmed to LRS. The initial state being HRS implies that the filament is ruptured. So switching to LRS means that filament must be formed between the top electrode (TE) and bottom electrode (BE). This is accomplished by applying a positive voltage between the top and bottom electrode. Since the bottom electrode of ReRAM is connected to the drain of the transistor, the source of the transistor is grounded, as depicted in Fig. 6.2. Now, V_{SET} appears across the composite 1T-1R device. The voltage V_G applied to the gate of the transistor switches ON the transistor to conduct current (it is assumed that $V_G > V_{TH}$, the threshold voltage of the NMOS transistor). The ON resistance of the transistor is very less compared to the HRS of the ReRAM cell and hence almost the whole of V_{SET} appears across the ReRAM cell. This results in the filament being formed in the ReRAM cell by redox reaction or electrochemical metallization. The formed filament forms a low resistive path between the top and bottom electrode, switching the device to LRS. The gate voltage decides the LRS to which the device is programmed. Hence, during SET process, the gate voltage can be varied to program the device to different LRS (elaborated more in Section 7.3.1).

To RESET the device from LRS to HRS, the polarity of the voltage applied across the ReRAM cell must be reversed *i.e.* the voltage must be applied to the bottom electrode of the ReRAM cell and the top electrode must be grounded. Since the bottom electrode is connected to the drain, V_{RESET} is applied to the source terminal of the access transistor and the top electrode is grounded, as depicted in Fig. 6.2. Now, V_{RESET} appears across the composite 1T-1R device. The filament is ruptured, switching the ReRAM cell to HRS. When the memory cells are configured into a 1T-1R array, the top electrodes of all the cells in a column are connected to the BL and the source terminals of all the cells in that column are connected to the SL (Fig. 6.2, bottom). During SET process, BL multiplexer selects one of the columns (according to the address) and connects it to V_{SET} while the SL multiplexer connects the corresponding SL to ground. WL goes high to select a particular row and the SET process is accomplished in the memory cell located in that row. Similarly, if RESET is intended in a cell, corresponding row is selected (by activating the WL) and the SL multiplexer selects the particular column and connects it to V_{RESET} while the BL multiplexer grounds the corresponding BL. The SET and RESET voltages can be generated on-chip from a band gap voltage reference as shown in Fig. 6.2. It must be noted that in some bipolar ReRAMs, the SET and RESET voltage are symmetrical *e.g.* V_{SET}= 1.2 V and V_{RESET}= -1.2 V. In this case, one set of resistors R_1, R_2 is enough.

6.2.1 Voltage Regulation during WRITE operation

An on-chip bandgap reference generator generates a precise voltage V_{REF}. The Low Dropout voltage regulator (LDO) is used to provide a constant voltage across the ReRAM cell (V_{SET}/V_{RESET}) in the presence of variations in the load current (current drawn by the 1T-1R cell). For example, during a SET operation, the ReRAM cell which is initially in HRS is switched to LRS, resulting in large variation in current drawn by the 1T-1R cell. The purpose of the LDO is to provide a constant V_{SET}/V_{RESET} across the programmed memory cell in the presence of such current varia-

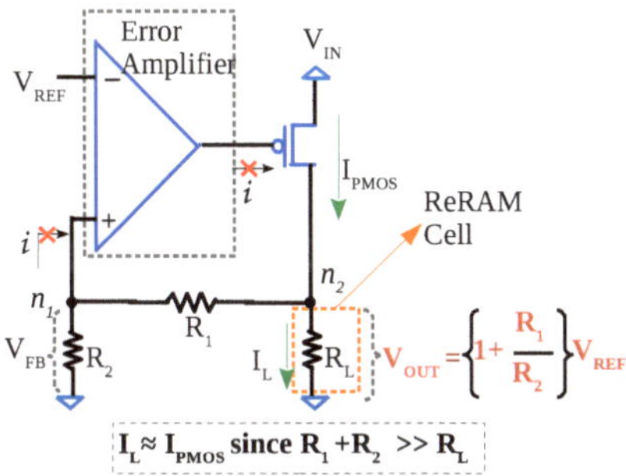

Fig. 6.3 The Low Dropout voltage regulator (LDO) is used to provide a constant voltage across the ReRAM cell (V_{SET}/V_{RESET}) in the presence of fluctuations in the load current (current drawn by the ReRAM cell) [4].

tion. Fig. 6.3 illustrates how the voltage is regulated using an LDO. Voltage regulation is achieved by feeding back a portion of the output load voltage, V_{OUT}. In this circuit, $R_1 + R_2 >> R_L$ so that most of the current through the PMOS flows into the load resistor, R_L (ReRAM cell in our case). Now if the current through R_L increases (as is the case during HRS $\rightarrow$ LRS transition), V_{OUT} increases ($V_{OUT} = I_L \cdot R_L$). A portion of V_{OUT} is fed back from the drain terminal of PMOS to the non-inverting input of the op-amp. This fed back voltage V_{FB} is $\frac{R_2}{R_1+R_2} \cdot V_{OUT}$ (since the current flowing into the op-amp is 0). Hence an increase in V_{OUT} results in a corresponding increase in V_{FB}. This increases the output of the error amplifier (op-amp). An increase in the gate voltage of a PMOS decreases the current through it and consequently decreases V_{OUT} ($I_{PMOS} \approx I_L$).

In this manner, the momentary increase in V_{OUT} due to the increase in load current is corrected and V_{OUT} remains constant. This intuitive explanation can be verified mathematically as follows. Since the op-amp is in a closed loop, it tends to pull the voltage at both its input terminals to be equal (ideal op-amp). This implies that the voltage at the non-inverting input is $\approx V_{REF}$. Current flowing from n_1 to ground through R_2 is (V_{REF}/R_2). Applying KCL at node n_1, this current could have come only from node n_2 since the current flowing into the op-amp is zero. Therefore, the current flowing from n_2 to n_1 through R_1 is also V_{REF}/R_2. The total voltage at node n_2 is

$$V_{OUT} = \frac{V_{REF}}{R_2} \times (R_1 + R_2) = \left(1 + \frac{R_1}{R_2}\right) V_{REF} \tag{6.1}$$

Thus, we observe that V_{OUT} is only a function of R_1, R_2 and V_{REF}, making the voltage across the R_L independent of V_{IN} and R_L. In this manner the voltage across R_L is regulated. R_1, R_2 and V_{REF} can be chosen such that $(1 + \frac{R_1}{R_2}) \cdot V_{REF}$ is equal to V_{SET}. The design of an LDO is non-trivial and reader is referred to [4, 5] for more details of the design of LDO.

6.3 Incremental Program and Verify Approach

We recall that variability is inevitable in ReRAM because the formation and rupture of conductive filament is inherently stochastic. This results in variability in the programmed states and is a great hindrance for its use as memory since variability affects reliability of memory operations. A great amount of research confirmed that programming the ReRAM cell by a single pulse (V_{SET}/V_{RESET} of 1 μs) results in an unreasonable amount of variability. To overcome the variability, two prominent approaches were pursued, the first being Incremental Program and Verify. The history dates back to the era of floating gate MOS transistors (flash memories) when storing more than two states was pioneered using this approach [6].

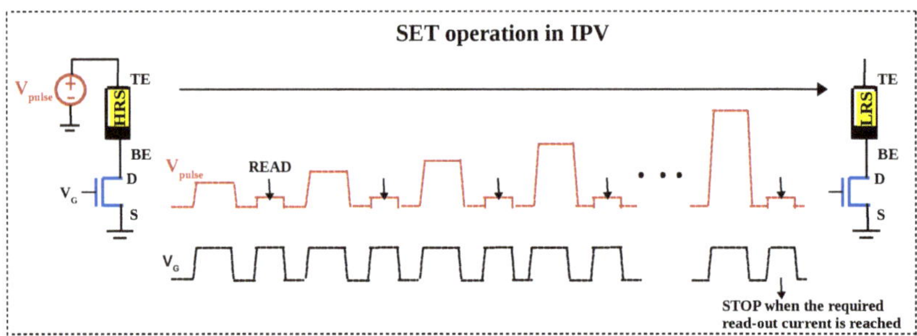

Fig. 6.4 In IPV, pulses of progressively increasing amplitude are applied and the programming operation is stopped once the target read-out current is achieved.

The basic idea is to program the ReRAM cell incrementally instead of a single pulse of amplitude V_{SET}/V_{RESET}. Suppose the statistical V_{SET} of a particular ReRAM technology has a mean of 1 V and a standard deviation of 0.2 V. This means that the cell can be switched by a pulse of amplitude 1±0.2 V. Instead of applying 1 V to switch the device from HRS to LRS, we apply a series of pulses starting from 0.6 V with 'verify' pulses in between. The 'verify' pulses are basically READ operations. But instead of converting the state to a logic voltage using sense amplifier, we read-out only the current from the memory cell. If the mean HRS and LRS are 500 KΩ and 10 KΩ, then with a 0.2 V READ voltage applied across the device, the current should be 0.4 μA and 20 μA, respectively. Hence the current state of the cell can be monitored by applying a 0.2 V pulse (verify) in between successive programming pulses, as depicted in Fig. 6.4. Pulses of progressively increasing amplitude are applied and the programming operation is stopped once the read-out current is 20 μA (corresponding to a LRS of 10 KΩ). In this manner, SET operation is achieved. The RESET operation (not depicted in Fig.6.4) can also be accomplished in a similar manner by applying the incremental pulses at the source of the transistor, while the TE is grounded. It must be noted that only the voltage across the 1T-1R device is increased in steps and the gate voltage of the transistor is held constant at a particular voltage [7].

Different variations of this IPV approach have been experimented and the reader is referred to [8, 9, 10, 11, 12]. For example, the width of the programming pulses have been varied and its effect on the programmed resistance is studied in [10]. Another variation pursued was to vary the gate voltage and the voltage at the top electrode was held constant during the SET process [12]. For better results using IPV, it is recommend that IPV approach be used in the electroforming stage of the ReRAM device before being used for the SET/RESET process [7].

6.4 WRITE Termination Approach

To overcome variations in the programmed state, the second approach pursued is the 'WRITE Termination' approach, also called self-termination or auto-termination. In this approach, instead of a series of pulses, a single pulse is used for programming the ReRAM cell. During the WRITE process (SET/RESET), the state of the cell is closely monitored. When the current through the cell reaches the target current, the writing process is terminated. In this section, we discuss two WRITE termination circuits from technical literature. The sequence of operation is same in both circuits: the current through the cell being written is monitored, the change in resistance sensed and the WRITE circuit disconnected from the ReRAM cell. In the first circuit, the BL current is monitored and in the second circuit, the SL current is monitored.

6.4.1 WRITE Termination by monitoring BL current

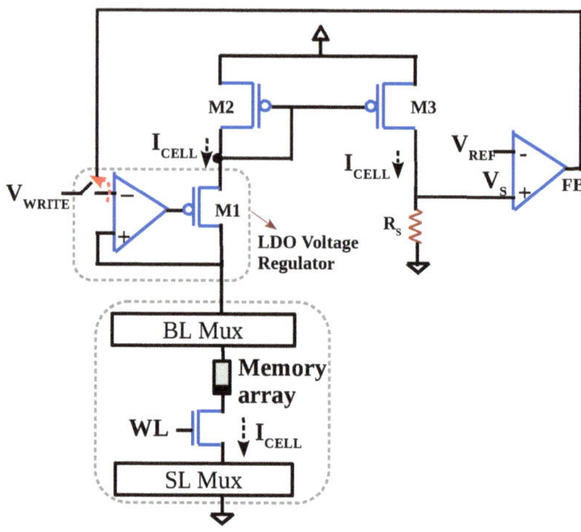

Fig. 6.5 WRITE Termination circuit during SET operation (as presented in [13] with minor modification)

The WRITE termination circuit (also called self adaptive WRITE Circuit) proposed in [13] is depicted in Fig. 6.5. The op-amp with a PMOS transistor M1 at its output acts as a voltage regulator and regulates the voltage V_{WRITE} applied across the cell (Notice that this is the same circuit presented in Fig. 6.3 with R_1=0 and $R_2 = \infty$. Consequently, the voltage at the drain of PMOS is simply V_{WRITE}). Let us assume that the ReRAM cell is at HRS. The circuit functions as a WRITE termination circuit during SET operation as follows. First, the SET operation is initiated by connecting V_{WRITE} to the inverting input of the op-amp. V_{WRITE} appears across the 1T-1R cell and the ReRAM device conducts a current I_{CELL} in proportion to its present resistance. The current drawn by the 1T-1R cell, I_{CELL} is mirrored by a PMOS current mirror. Transistors M2-M3 form a PMOS current mirror and the mirrored current I_{CELL} flows through a resistor R_S. In this manner, the current drawn by the 1T-1R cell is converted to an equivalent voltage, V_S. V_S and V_{REF} are compared using a simple voltage comparator. Resistor R_S is designed in such a way that $V_S = I_{CELL} \times R_S$ is less than V_{REF} when I_{CELL} is below a particular threshold. That threshold current I_{th} is the current through the cell when the switching process is considered complete. During SET operation, the I_{CELL} increases from a low current (HRS) till I_{th}. During this time, V_S is less than V_{REF} and the feedback signal FB is low. As soon as I_{CELL} increases above I_{th}, V_S becomes greater then V_{REF} and this makes FB go high. The high in FB is used as a trigger signal to disconnect the op-amp from the V_{WRITE}, terminating the WRITE operation. In this manner, by continuously monitoring the ReRAM cell current, WRITE termination is achieved. Other variations and improvements to this circuit can be found in [14, 15]

6.4.2 WRITE Termination by monitoring *SL* current

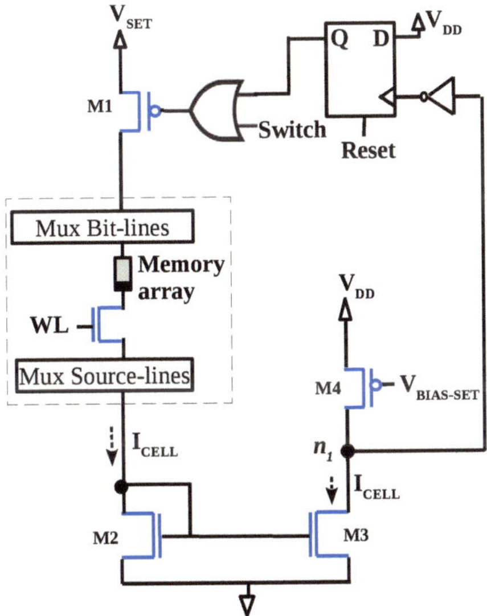

Fig. 6.6 WRITE termination circuit during SET process (as presented in [16] with minor modification)

The second WRITE termination circuit is presented in [16]. Fig. 6.6 depicts the WRITE termination process used during the SET process. Initially, the D flip-flop is reset so that Q is low (0 volts). The 'switch' signal is a signal to signify the start of the SET process and is active low, *i.e.* a low in 'switch' signifies the start of the SET process. Since 'switch' and Q are both low, the PMOS transistor M1 is ON and V_{SET} is connected to the Bit-line and therefore V_{SET} is connected to the top electrode of the 1T-1R cell. This initiates the SET process and the device begins to switch from HRS to LRS. Consequently, the current through the memory cell, I_{CELL} starts to increase. Transistors M2-M3 form an NMOS current mirror and therefore I_{CELL} starts to discharge node n_1. It must be noted that node n_1 is maintained at a pre-defined potential near V_{DD} by $V_{BIAS-SET}$. In other words, when I_{CELL} is low (HRS), it is not strong enough to pull node n_1 below $\frac{V_{DD}}{2}$. This implies that n_1 is almost at V_{DD} and hence clock of the flip-flop is low. However, this state of equilibrium will be disrupted by the switching of the ReRAM cell from HRS to LRS. As the state of the ReRAM cell transitions to LRS, there will be an increase in I_{CELL} which is copied by the current mirror. Consequently, node n_1 falls below $\frac{V_{DD}}{2}$ and triggers a positive edge (low to high transition) at the clock pin of the D-flip flop. This positive edge makes Q go high (since D is high at the positive edge of the clock) and the output of the OR gate goes high. This switches OFF the PMOS transistor M1 and V_{SET} is electrically disconnected from the 1T-1R cell. The SET process is terminated.

$V_{BIAS-SET}$ is carefully chosen so that clock pin goes high as soon as the cell current increases above a target threshold current, I_{th}. In other words, $V_{BIAS-SET}$ is chosen such that node n_1 is maintained above $\frac{V_{DD}}{2}$ for currents below I_{th}. In this manner, as soon as the cell current rises above I_{th}, n_1 goes below $\frac{V_{DD}}{2}$ and the clock pin of D-FF goes high, terminating the SET operation. This WRITE termination circuit, as presented in [16] has a limitation. Out of the V_{SET} applied to the 1T-1R cell, there is a voltage drop across the M2 transistor since the SL is not grounded but connected to the current mirror formed by M2-M3 ($V_{SET} = V_{1T-1R} + V_{DS}$ of M2). This is because the drain of M2 must be maintained at a voltage that can at least switch ON transistors M2 and M3 (V_{GS} of M2,M3 must be larger than Vth, the threshold voltage of NMOS). This complicates the peripheral circuitry since the SL, which is usually grounded during SET operation must be connected to another circuitry. In contrast, the WRITE termination approach of [13] requires only a modification of the peripheral circuitry at the BL side during SET operation.

It must be noted that ReRAM cells exhibit wide variation in their switching time, *i.e.* t_{SET}, the time a particular cell takes to switch from HRS to LRS varies from cycle-to-cycle and also across cells in an array. In such cases, this WRITE termination circuit is very valuable since at the end of the SET process, V_{SET} is disconnected from the 1T-1R cell and does not waste energy. Many similar circuits with the aim of terminating the WRITE process have been reported. The reader is referred to [17, 18] for variations and improvements to this write termination approach. In addition to minimizing variations of the programmed states, the write termination approach also aids in minimizing power consumed by the WRITE process [16].

6.5 Writing into 1S-1R arrays

So far, we considered writing into 1T-1R arrays. As stated, 1T-1R avoids sneak currents and enables efficient programming, *i.e.* the voltage applied across the cell reaches only cell intended to be written and there is no energy wastage. When there is no access transistor, writing becomes a bit complicated. The reader might recall that to achieve dense storage, ReRAM is also fabricated in 1S-1R structure. Instead of an access transistor for each cell, a selector device is integrated in series with the ReRAM cell (Section 4.2.2). The purpose of the selector device is to minimize the effect of sneak currents. While writing into a particular cell, the adjacent cells are carefully biased at specific voltages to avoid the disturbance of their states.

Unlike a 1T-1R array, the 1S-1R array has both the terminals of the ReRAM cell shared with adjacent cells in the same row/column. As depicted in Fig. 6.7-(a), cells 1,4,7 along the same row share the WL_1. Similarly, cells 4,5,6 along the same column share BL_2. This sharing of WL/BL between adjacent cells results in an unnecessary voltage across cells where WRITE operation is not intended. Consider the case when we want to write into cell 9. Then V_W is applied to WL_3 and BL_3 can be grounded resulting in a voltage of V_W across the 1S-1R cell (we assume a bipolar ReRAM cell with symmetrical switching voltages *i.e.* $V_{SET} = |V_{RESET}| = V_W$). Let us assume that all other WLs and BLs are grounded since no operation is intended in other rows/columns. As depicted in Fig. 6.7-(a), this results in a voltage of V_W across cells 3 and 6. Although, our intention was to write into cell 9, cells 3 and 6 will also be written since

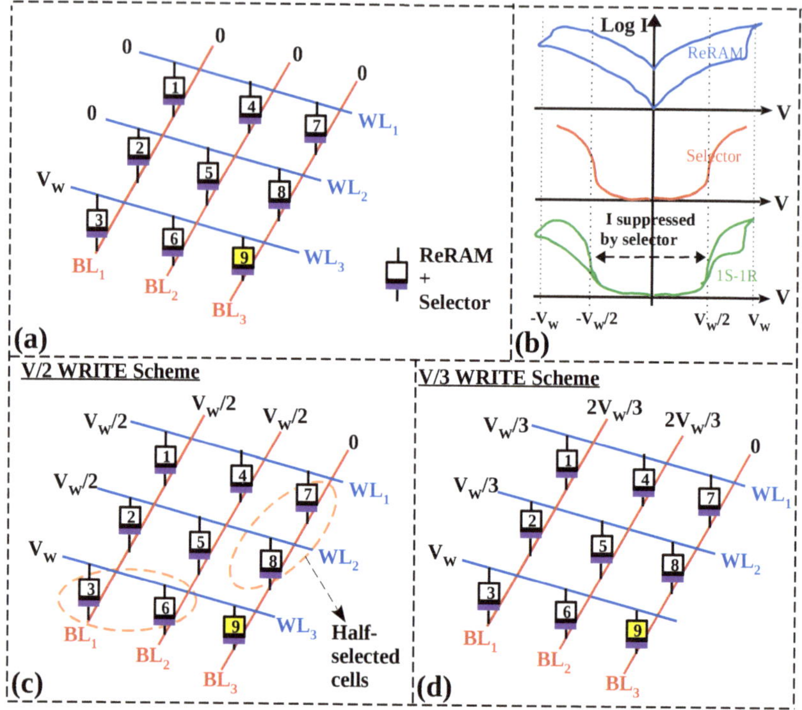

Fig. 6.7 For writing into 1S-1R arrays, special bias schemes like $V/2$ and $V/3$ scheme are needed to avoid disturbing the state of unselected cells.

they have V_W across them. In other words, it is not possible to write into cell 9 without disturbing the data in cells 3 and 6. All other cells have 0 voltage across them and there is no danger of their state getting disturbed (Fig. 6.7 (a)). Note that this was not a problem when writing into 1T-1R cells because of the access transistor in each cell *i.e.* all $WL/BL/SL$ where WRITE operation is not intended can be grounded and it does not disturb the stored state.

To overcome this problem, two writing schemes have been popularly used: $V/2$ and $V/3$ scheme. As depicted in Fig. 6.7-(c), the WRITE voltage V_W is applied across cell 9 by biasing WL_3 at V_W and BL_3 at 0 V. As the name suggests, in the $V/2$ scheme, all other WL/BLs are biased at $V_W/2$. Notice that cells 3,6,7,8 have $V_W/2$ across them. Hence, they are called half-selected cells. The data in half-selected cells will not be disturbed during write operation since they have half the SET voltage across them. As illustrated in Fig. 6.7-(b), at half the SET voltage ($V_W/2$), the current through the 1S-1R cell is highly reduced, resulting in less leakage currents through half-selected cells during WRITE operation. Cells 1,2,4,5 have 0 V across them (Fig. 6.7-(c)) and there is no danger of their states getting disturbed. To perform RESET operation on cell 9, - V_W is applied across the cell by applying V_W to BL_3 and WL_3 is grounded.

In $V/3$ WRITE scheme, except the row/column being written, all WLs are biased at $V_W/3$ and all BLs are biased at $2.V_W/3$ for SET operation (Fig. 6.7-(d)). The reader can verify that, except cell 9 which has V_W across it, all other cells have $|V_W/3|$ across them. Hence, there is no danger of the other cells being written or disturbed. Since the cells where WRITE operation is not intended are biased at $|V_W/3|$, the currents through them is negligible (suppressed by the selector), as shown in Fig. 6.7-(b). Similarly, to perform RESET operation on cell 9, V_W is applied to BL_3 and WL_3 is grounded. All other BLs are biased at $V_W/3$ and all other WLs are biased at $2.V_W/3$.

The following is a summary of the benefits and drawbacks of these two write schemes: Generally speaking, the V/2 WRITE scheme consumes less power than the V/3 scheme. This is due to the fact that all of the unselected cells in the V/3 scheme have $V_W/3$ across them, consuming static power during the WRITE operation. On the contrary, the unselected cells in the V/2 scheme (not along the chosen WL and BL) have zero voltage across them ideally. However, the maximum voltage that the unselected cells have across them in the V/3 scheme is $V_W/3$, but in the V/2 scheme it is $V_W/2$. This means that the V/3 method is more immune to write disturbances than the V/2 scheme [19].

6.6 Issues faced while writing into ReRAM cell

6.6.1 Write disturb

Although the $V/2$ and $V/3$ WRITE schemes ensure a maximum of $V_W/2$ and $V_W/3$ across the half selected/unselected cells, cell's state have been disturbed in experiments. 'Write disturb' has also been called 'Programming disturb' in some publications since this disturb happens when we are programming the cells in the 1S-1R array. The reason for the change in the stored state is a cumulative effect *i.e.* although a single $V_W/2$ pulse across an half-selected cell cannot disturb its state, a series of such pulses is found to disturb its state [20]. During consecutive WRITE operations, half-selected cells are subject to a series of $V_W/2$ voltage stress across them in $V/2$ WRITE scheme. According to the detailed study undertaken in [21, 20], write disturb occurs due to additional generation and recombination of oxygen vacancies and ions under voltage stress on the half-selected cells. In simple terms, HRS disturb (a half-selected cell in HRS getting disturbed to a LRS) is due to a progressive growth of the conductive filament due to many $V_W/2$ pulses applied. In the array-level simulation, it was observed that half-selected cells closer to the voltage source suffered from more serious write disturbs. Furthermore, the write disturb becomes more serious as the array size increases [20].

6.6.2 IR drop

Strictly speaking IR drop is a characteristic of the array, but we discuss it under this section because the effect of IR drop is more profound when writing into a ReRAM array. As technology scales down, the interconnect resistance increases. The IR drop problem becomes severe when the WL/BL wire widths scales to sub-50 nm regime where the interconnect resistivity drastically increases due to the increased electron surface scattering [19]. For example, at 20-nm node, the copper interconnect resistance between adjacent cells is $\approx 2.93\ \Omega$; thus, the IR drop along the wire for a large array cannot be neglected. For a 1024×1024 array, such an interconnect resistance will result in the cell farthest from the WL driver seeing an interconnect resistance $\approx 3k\Omega$. If the ReRAM cell's LRS resistance (typically a few kΩ up to tens of kΩ) is comparable to this interconnect resistance, a portion of the write voltage will drop on the wire instead of the ReRAM cell. To achieve a successful write operation, the write voltage from the driver has to be increased above the actual switching voltage of the ReRAM cell to compensate for the IR drop. However, the write voltage cannot be increased much since doing so will disturb the state of the half-selected cells close to the driver.

References

[1] E. Ambrosi, A. Bricalli, M. Laudato, and D. Ielmini, "Impact of oxide and electrode materials on the switching characteristics of oxide reram devices,"

Faraday Discuss., vol. 213, pp. 87–98, 2019.

[2] J. Reuben, M. Biglari, and D. Fey, "Incorporating variability of resistive ram in circuit simulations using the stanford–pku model," *IEEE Transactions on Nanotechnology*, vol. 19, pp. 508–518, 2020.

[3] S.-S. Sheu, K.-H. Cheng, M.-F. Chang, P.-C. Chiang, W.-P. Lin, H.-Y. Lee, P.-S. Chen, Y.-S. Chen, T.-Y. Wu, F. T. Chen, K.-L. Su, M.-J. Kao, and M.-J. Tsai, "Fast-write resistive ram (rram) for embedded applications," *IEEE Design & Test of Computers*, vol. 28, no. 1, pp. 64–71, 2011.

[4] B. Razavi, "The low dropout regulator [a circuit for all seasons]," *IEEE Solid-State Circuits Magazine*, vol. 11, no. 2, pp. 8–13, 2019.

[5] B. Razavi, "The design of an ldo regulator [the analog mind]," *IEEE Solid-State Circuits Magazine*, vol. 14, no. 2, pp. 7–17, 2022.

[6] R. Bez, E. Camerlenghi, A. Modelli, and A. Visconti, "Introduction to flash memory," *Proceedings of the IEEE*, vol. 91, no. 4, pp. 489–502, 2003.

[7] E. Pérez, A. Grossi, C. Zambelli, P. Olivo, and C. Wenger, "Impact of the incremental programming algorithm on the filament conduction in hfo2-based rram arrays," *IEEE Journal of the Electron Devices Society*, vol. 5, no. 1, pp. 64–68, 2017.

[8] V. Milo, A. Glukhov, E. Pérez, C. Zambelli, N. Lepri, M. K. Mahadevaiah, E. P.-B. Quesada, P. Olivo, C. Wenger, and D. Ielmini, "Accurate program/verify schemes of resistive switching memory (rram) for in-memory neural network circuits," *IEEE Transactions on Electron Devices*, vol. 68, no. 8, pp. 3832–3837, 2021.

[9] D. M. Nminibapiel, D. Veksler, P. R. Shrestha, J. P. Campbell, J. T. Ryan, H. Baumgart, and K. P. Cheung, "Impact of rram read fluctuations on the program-verify approach," *IEEE Electron Device Letters*, vol. 38, no. 6, pp. 736–739, 2017.

[10] E. Perez, O. Gonzalez Ossorio, S. Duenas, H. Castan, H. Garcia, and C. Wenger, "Programming pulse width assessment for reliable and low-energy endurance performance in al:hfo2-based rram arrays," *Electronics*, vol. 9, no. 5, 2020.

[11] J.-C. Liu, T.-Y. Wu, and T.-H. Hou, "Optimizing incremental step pulse programming for rram through device–circuit co-design," *IEEE Transactions on Circuits and Systems II: Express Briefs*, vol. 65, no. 5, pp. 617–621, 2018.

[12] V. Milo, F. Anzalone, C. Zambelli, E. Perez, M. K. Mahadevaiah, O. G. Ossorio, P. Olivo, C. Wenger, and D. Ielmini, "Optimized programming algorithms for multilevel rram in hardware neural networks," in *2021 IEEE International Reliability Physics Symposium (IRPS)*, pp. 1–6, 2021.

[13] X. Xue, W. Jian, J. Yang, F. Xiao, G. Chen, S. Xu, Y. Xie, Y. Lin, R. Huang, Q. Zou, and J. Wu, "A 0.13 µm 8 mb logic-based cu $_x$si $_y$o reram with self-adaptive operation for yield enhancement and power reduction," *IEEE Journal of Solid-State Circuits*, vol. 48, no. 5, pp. 1315–1322, 2013.

[14] C.-F. Lee, H.-J. Lin, C.-W. Lien, Y.-D. Chih, and J. Chang, "A 1.4mb 40-nm embedded reram macro with 0.07um2 bit cell, 2.7ma/100mhz low-power read and hybrid write verify for high endurance application," in *2017 IEEE Asian Solid-State Circuits Conference (A-SSCC)*, pp. 9–12, 2017.

[15] Y. L. Song, Y. Meng, X. Y. Xue, F. J. Xiao, Y. Liu, B. Chen, Y. Y. Lin, Q. T. Zou, R. Huang, and J. G. Wu, "Reliability significant improvement of

resistive switching memory by dynamic self-adaptive write method," in *2013 Symposium on VLSI Technology*, pp. T102–T103, 2013.

[16] M. Alayan, E. Muhr, A. Levisse, M. Bocquet, M. Moreau, E. Nowak, G. Molas, E. Vianello, and J. M. Portal, "Switching event detection and self-termination programming circuit for energy efficient reram memory arrays," *IEEE Transactions on Circuits and Systems II: Express Briefs*, vol. 66, no. 5, pp. 748–752, 2019.

[17] W.-H. Chen, W.-J. Lin, L.-Y. Lai, S. Li, C.-H. Hsu, H.-T. Lin, H.-Y. Lee, J.-W. Su, Y. Xie, S.-S. Sheu, and M.-F. Chang, "A 16mb dual-mode reram macro with sub-14ns computing-in-memory and memory functions enabled by self-write termination scheme," in *2017 IEEE International Electron Devices Meeting (IEDM)*, pp. 28.2.1–28.2.4, 2017.

[18] Z. Wang, Z. He, R. Yang, S. Fan, J. Lin, F. Liu, Y. Jia, C. Yuan, Q. Tang, and L. Jiang, "Self-terminating write of multi-level cell reram for efficient neuromorphic computing," in *2022 Design, Automation & Test in Europe Conference & Exhibition (DATE)*, pp. 1251–1256, 2022.

[19] S. Yu and P.-Y. Chen, "Emerging memory technologies: Recent trends and prospects," *IEEE Solid-State Circuits Magazine*, vol. 8, no. 2, pp. 43–56, 2016.

[20] Z. Chen, H. Li, H.-Y. Chen, B. Chen, R. Liu, P. Huang, F. Zhang, Z. Jiang, H. Ye, B. Gao, L. Liu, X. Liu, J. Kang, H.-S. P. Wong, and S. Yu, "Disturbance characteristics of half-selected cells in a cross-point resistive switching memory array," *Nanotechnology*, vol. 27, p. 215204, apr 2016.

[21] H. Li, H.-Y. Chen, Z. Chen, B. Chen, R. Liu, G. Qiu, P. Huang, F. Zhang, Z. Jiang, B. Gao, L. Liu, X. Liu, S. Yu, H.-S. P. Wong, and J. Kang, "Write disturb analyses on half-selected cells of cross-point rram arrays," in *2014 IEEE International Reliability Physics Symposium*, pp. MY.3.1–MY.3.4, 2014.

Chapter 7
Row/Column Access Circuits and Multilevel Cell

7.1 Row Decoder: Circuit to Select a Row

The function of the row-decoder is to select a row, which needs to be written or read from. Electrically, this is accomplished by driving the row wire to a high voltage. In ReRAM, assuming 1T–1R configuration, selecting a row means that we need to make one particular WL out of n WLs go high. One particular WL being high makes the access transistor to be ON, enabling the flow of current through the ReRAM cell for reading/writing. Conventionally, row decoders accomplish this by a combinatorial logic. While designing a row decoder, it is important to minimize the latency/delay of the decoding process. From the system point of view, for a memory technology to be competitive, the memory access time, *i.e.* the time required to read (which includes the time to select a row) from it must be minimum so that the memory access does not slow down the entire system. Conventionally, row decoders were designed using Boolean gates as a combinational logic. A typical decoder has n inputs and 2^n outputs and for each input combination, only one of the 2^n output signals goes high. Fig. 7.1 illustrates an example of 2:4 decoding using the common types of decoders – static NAND, dynamic NAND and dynamic NOR decoders.

Static NAND decoder simply uses a CMOS NAND gate for each row. Depending on the address (A_1A_0), only one of the four outputs $X_3X_2X_1X_0$ will be LOW and all others will be HIGH. For n-inputs, the decoder will occupy huge area since each row needs an n-input NAND gate. The area will be extremely large as n grows. As depicted in Fig. 7.1, dynamic decoders have a single PMOS transistor in the pull-up network of

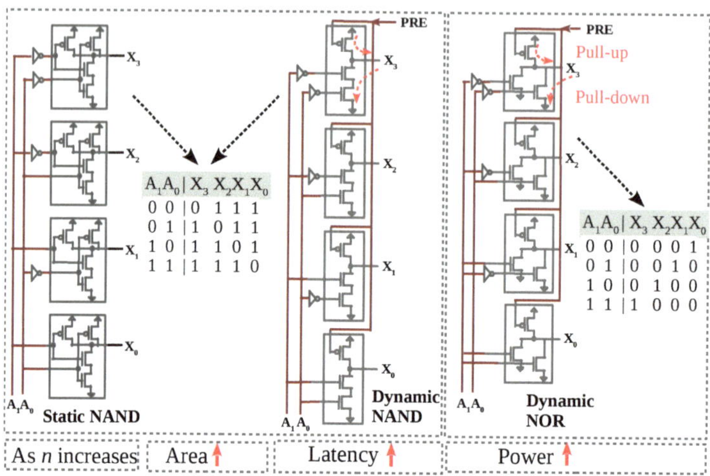

Fig. 7.1 Example of 2:4 decoding using static NAND, dynamic NAND and dynamic NOR decoders. (A_1A_0) represent the input address and ($X_3X_2X_1X_0$) are the four outputs which are connected to the four rows/ WLs.

each row (thus offering area reduction compared to static decoders). PRE signal is initially LOW and this pre-charges the output node of the decoder to V_{DD}. During the actual decoding, PRE goes high and depending on the inputs (A_1A_0), one of the four outputs is selected. In dynamic NAND decoder, the pull-down network is similar to a CMOS NAND gate (hence the name). Depending on the inputs, one of the four outputs will be pulled down and all other outputs are high. The dynamic NOR decoder has a pull-down network similar to a CMOS NOR gate. It can be verified that for a dynamic NOR decoder, only one of the four outputs remain HIGH and all other outputs are pulled down to ground during decoding. This implies that for a 8:256 dynamic NOR decoder, 255 nodes are pulled down to ground during decoding process. In contrast, for the same decoding task, only a single output node is pulled to ground during NAND decoding. Hence dynamic NOR decoders consume more power when compared to dynamic NOR decoders [1]. However, dynamic NAND decoders are slower than dynamic NOR decoders because the output node is discharged through stacked transistors (as n increases, the output node is discharged through n transistors while it is discharged through a single transistor in dynamic NOR).

Among these types of decoders, dynamic NAND decoder is one feasible row decoder adopted for ReRAM array due to its compact area

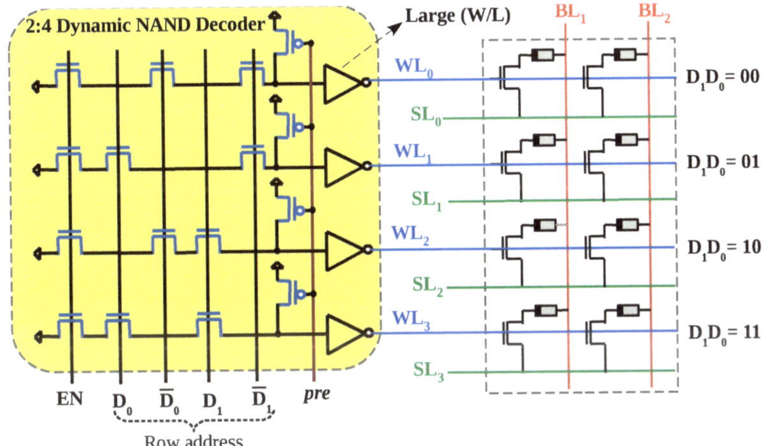

Fig. 7.2 A dynamic NAND decoder uses a pre-charge signal 'pre', which when low, all WLs are driven to '0'. When 'pre' goes high, WL_i corresponding to D_1D_0 goes high provided 'EN' is high. The value of the D_1D_0 needed to select a particular row is shown on the right.

[2]. As depicted in Fig.7.2, the dynamic decoder uses two transistors which select one out of four rows and a third transistor in each row, whose gate is connected to 'EN' signal. When the row decoder needs to be deactivated, the pre-charge signal 'pre' is made low. In this case, the input of all the inverters will be pulled up to V_{DD} and consequently all the WLs will be low. Hence, all transistors will be OFF and there will no current flow in the entire memory array. When 'pre' goes high, the row decoder is active and is ready to select one of the four rows. Note that 'pre' being high implies that all the PMOS transistors are OFF. As soon 'pre' goes high, 'EN' signal also goes high, signalling the start of the row decoding process. The reader is reminded that the input nodes of the inverters in each row were pulled up to V_{DD} previously. Depending on D_1D_0, one of the four rows will be pulled to ground. *e.g.* for D_1='0' and D_0 ='1', only the input of the inverter in the second row will be pulled to ground (since 'EN' is also high). Consequently, WL_1 will be high and all other WLs will be low. In this manner, a 2:4 decoder selects one of the four rows corresponding to D_1D_0 which acts as the row address.

In Fig.7.2, the source of the PMOS transistors in each row is shown connected to V_{DD}. So each WL will be raised to V_{DD} if selected and the unselected rows will be at ground potential. Usually, a voltage of V_{DD} is enough to switch ON the access transistor for READ/WRITE operations. In 130 nm process node, V_{DD} is 1.2 V and a gate voltage of 1.2 V is enough for READ/WRITE process of 1T-1R cell. In some ReRAM technology, it is possible that a ReRAM cell requires a voltage greater than V_{DD} at the gate of the access transistor during SET/RESET process. In this case, level shifter circuits may be necessary in each row since by default, the row decoder pulls up the WL of the selected row to V_{DD} only.

7.1.1 WL driver

Some books distinguish between Row Decoder and Word Line Driver (WL Driver) as two separate circuits. Strictly speaking, these are two different functions. Row decoder is used to only select a row and WL driver is used to drive the WL to a particular voltage. The function of the WL driver is to make the WL (when selected) rise from 0 V to a stable voltage. This involves charging the large parasitic capacitance – the gate capacitance of the access transistors and the wire capacitance. This is accomplished by making the transistor driving the WL large (for good current driving capability). In many cases, such a driving capability can be integrated to the row decoder itself by making the output transistors of the row decoder large. The inverters which drive the WLs (Fig. 7.2) should be sized in such a manner that they are able to drive the WL wire capacitance and the gate capacitance of the access transistor.

7.2 Column Multiplexers

The function of the column multiplexer is to select a column and connect it to the READ/WRITE circuitry. Unlike the row decoder, the column multiplexer does not need to drive the BL/SL to a particular voltage. Hence, column multiplexers are basically routing switches which take a column address as the input, select the corresponding BL/SL and connect it to the READ/WRITE circuitry (it is the function of the READ/WRITE circuitry to drive the BL/SL to a particular voltage). However, since the column multiplexer also takes a column address as input and does some decoding, it is common to refer to column multiplexers as column decoders. Some books simply refer to them as column control circuitry since these circuits are used to control a column – selecting (addressing) and connecting it to the required READ/WRITE circuit to achieve the required memory operation.

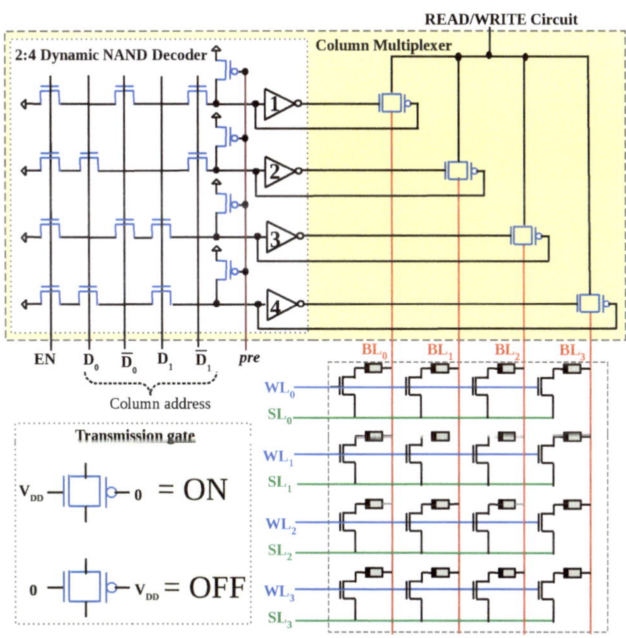

Fig. 7.3 Using a dynamic decoder, the column multiplexer selects a specific column and electrically connects it to READ/WRITE circuit through a transmission gate.

Fig. 7.3 depicts a simple column multiplexer. The column multiplexer consists of a dynamic decoder and transmission gates connected to each of the four outputs of the decoder. The 2:4 dynamic decoder is used as an address decoder to decode the column address. Like the row decoder discussed in previous section, the column address D_1D_0 is used to select a specific column out of the four columns. Let us assume we intend to write into the second column from left. This means that BL_1 alone must be connected to the READ/WRITE circuit. When 'pre' signal is low, input of all the four inverters are high and hence all the transmission gates which are connected to the output of the inverters are OFF. To select a column, both 'pre' and 'EN' should go high. When 'pre' is high, all the PMOS transistors are OFF and hence the capacitive charge at the input of the inverters is free to discharge, if there is a path. If D_1 =0 and D_0 =1, the input of inverter 2 discharges to ground since there is a path. The input of inverters 1,3,4 continue to be high since they do not have a path to discharge. Therefore, only the transmission gate in the second row is ON. In this manner, only BL_1 is connected to the READ/WRITE circuit and other BLs are not connected. The transmission gates need to be sized such that they can conduct enough current (the current flowing through the 1T-1R cell during WRITE process).

Such a column control circuitry is very beneficial from the layout/area point of view. The selection transmission gates can be laid out along each column and the whole multiplexer is area efficient. For 1T-1R configuration, a similar column multiplexer is needed at the end of the SLs to select one of them and connect them to the READ/WRITE circuit (not shown in Fig. 7.3). This is because, in 1T-1R configuration, READ/WRITE operations are performed by applying voltages between the top electrode (BL) and the source of the 1T-1R transistor (SL).

7.2.1 Predecoding for large arrays

The decoders we have presented in Fig. 7.1 are called single-stage decoders (n-to-2^n). For increasing array size, row/column decoding is performed usually in multiple stages [3]. This is because, as n (number of rows to be selected) increases, the size of the individual combinational logic in each row increases making the decoder large and also affects the decoding speed. Fig. 7.4 illustrates how a 6:64 decoder is implemented using three 2:4 decoder followed by AND gates. In addition to reducing access time, predecoding also enables pitch-matching *i.e.* the 64 AND gates of the final decoder are laid out along the memory array and the predecoder can be laid out farther away from the array. If the same 6:64 decoding was accomplished by single stage decoding, we would have required a six-input NAND gate instead of the three-input AND gate near the array. By predecoding, we have reduced the number of stacked transistors from six to three for NAND decoder. For n greater than 5, it is advisable to use a pre-decoder (*i.e.* multiple stages instead of single-stage decoding) [4].

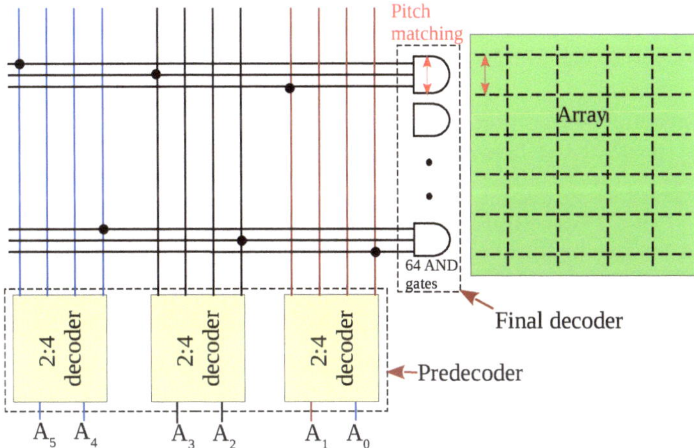

Fig. 7.4 Illustration of a 6:64 decoding using predecoding technique. Predecoding enables reduction of decoding latency and an area-efficient layout [1].

7.3 Multi-Level Cell(MLC): Some Directions for Writing and Reading

Similar to flash memory technology, storing more than two states in ReRAM cell was also pursued in ReRAM technology. The obvious advantage of storing more than two states was the increase in memory density. Furthermore, many novel applications of ReRAM like matrix vector multiplication (described in the next chapter) require programming the cell to more than two states. Therefore, in this section, we give a brief overview how the ReRAM cell can be programmed to multiple states and also how multiple states can be distinguished using appropriate Sense Amplifiers.

7.3.1 Writing Multiple States

As stated in Section 4.4.2, ReRAMs can be programmed to multiple states in following ways [5]:

1. By varying compliance current (also called multilevel SET): from the initial HRS, the ReRAM is programmed to different LRS during the SET process.
2. By varying reset voltage (also called multilevel RESET): from the initial LRS, the ReRAM is programmed to different HRS during the RESET process.
3. By varying programming pulse widths (not popularly pursued due to being energy inefficient).

It has been difficult to achieve good, reliable multiple states in ReRAM since the variability problem which existed while programming the cell to two states persists while programming the cell to multiple states. Even if a certain level of programming accuracy was achieved at the device level (single-cell), it could not be replicated at the array level (*i.e.* across many cells). However, among the three aforementioned methods, multilevel SET process has shown the most reliability at the array level [6]. Therefore, in this section, we will briefly describe the multilevel SET process.

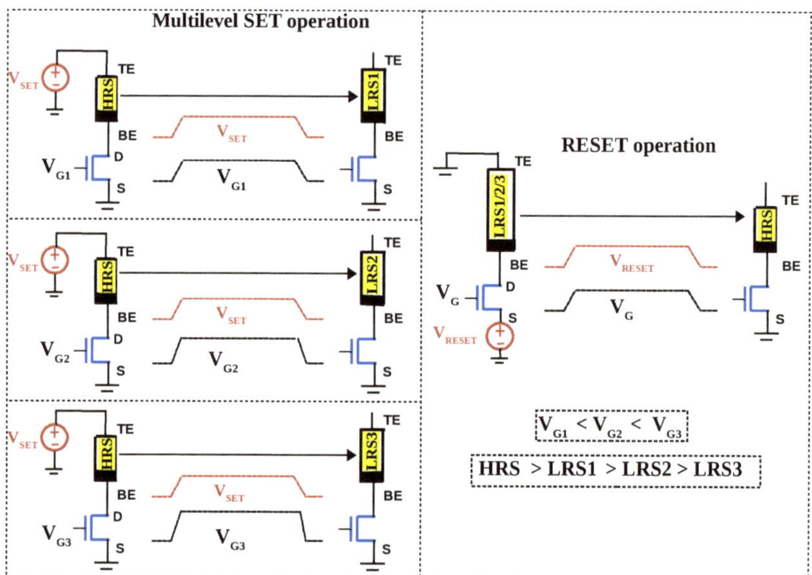

Fig. 7.5 The device is programmed to different LRS by varying the gate voltage during the SET process, while the voltage applied to the top electrode remains the same. Higher the gate voltage, lower the programmed resistance.

As depicted in Fig. 7.5, in multilevel SET process, the device is programmed to different LRS during the SET process, *i.e.* HRS to LRS transition. This implies that if we want to accommodate four states in the device, we need to be able to program the device to three different LRS (the fourth state being the HRS). Different LRS states are achieved by varying the gate voltage of the access transistor during SET process. In ReRAM, the resistance of the state is decided by the thickness of the filament. As the filament becomes thicker, the resistance of the ReRAM cell decreases. A higher gate voltage will enable more current to flow through the 1T-1R cell and, consequently forms a thicker filament in the ReRAM device. LRS3 is the lowest resistance since V_{G3} is the highest gate voltage. As illustrated in Fig. 7.5, V_{G1}, V_{G2}, V_{G3} at the gate of the access transistor results in LRS1,LRS2,LRS3 respectively. For example, the 1T–1R cell of IHP with a HRS of 66.6 KΩ can be programmed to LRS of 10KΩ, 6.6 KΩ and 5 KΩ by applying a gate voltage of 1.2 V, 1.4 V and 1.6 V, respectively [7]. It must be observed that although the device is getting programmed to different LRS, the RESET process remains the same. Irrespective of the LRS, the device gets re-programmed to HRS with a particular gate voltage V_G and the voltage applied to the source of the transistor. This is because, during RESET, the filament is broken.

7.3.2 Reading Multiple States

Reading more than two states requires using the same principle of reading two states (discussed in Chapter 5), with extra circuits to accommodate the reading of the additional states. A Sense Amplifier (SA) is basically an Analog-to-Digital Converter (ADC). It must convert the analog resistance of the memory cell to a voltage of V_{DD} or 0 volts (logic '1' or '0'). For two states, this was simple and one of the modes like voltage-mode or current-mode was used to convert the resistance of the cell to an equivalent voltage or current and then compared with a reference quantity, V_{REF}/I_{REF}. However, after the state of the cell is converted to an equivalent voltage/current, we needed to compare with only a single reference quantity in two state sensing. For sensing more than two states, a single reference is not enough.

Conventionally, there were two methods for sensing more than two states: sequential and parallel. Both were developed to read data out

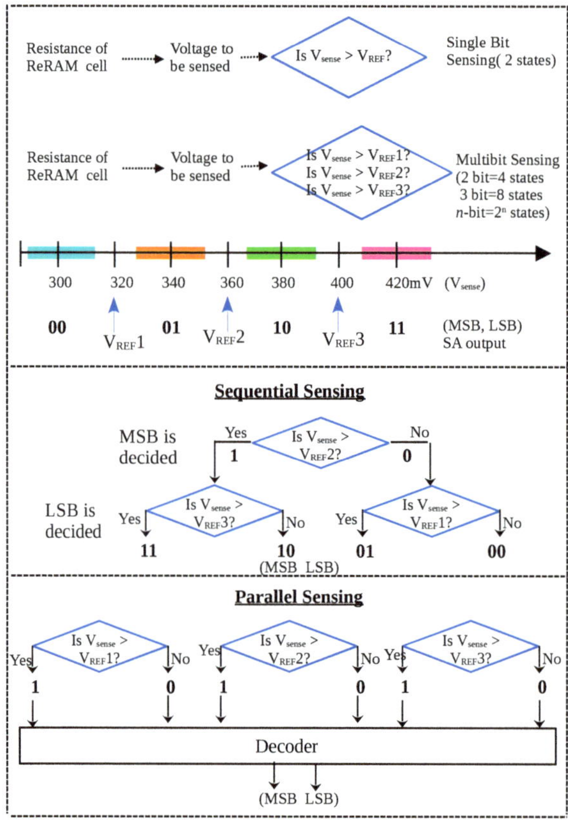

Fig. 7.6 Similar to two-state sensing, resistance of the ReRAM cell is converted to an voltage and compared with a reference voltage in multi-level sensing. Multiple comparisons are required while sensing multiple states

of MLC flash memory. In the sequential approach to multi-level sensing, a single SA is used and numerous comparisons are performed by sequentially changing the reference quantity (voltage or current), resulting in the identification of the cell resistance [8]. The parallel approach uses numerous sense amplifiers and compares the read quantity with the reference quantity simultaneously, similar to a flash ADC [9]. The former approach has lower hardware complexity but high latency, while the latter approach achieves high measurement speed at the expense of hardware. Fig. 7.6 illustrates both these types. The sequential sensing illustrated in Fig. 7.6 is also called dichotomic (binary search) approach and in this methodology, the MSB is first decided by comparing V_{sense} with the center V_{REF} and based on the MSB, one of the other V_{REF}s is used to decide the LSB [8, 10]. Assuming the voltage to be sensed varies from 300 to 420 mV (Fig. 7.6), in sequential sensing, a center V_{REF} of 360 mV is compared with V_{sense} (voltage to be sensed) to decide the MSB. If MSB is 1, a higher V_{REF} of 400 mV is used to decide the LSB. On the contrary, if MSB is 0, a lower V_{REF} of 320 mV is used to decide the LSB. In parallel sensing, V_{sense} is compared with three V_{REF}s simultaneously and their outputs are decoded to two bits. The comparison between V_{sense} and V_{REF}s can be implemented using the voltage comparators discussed in Section 5.3.

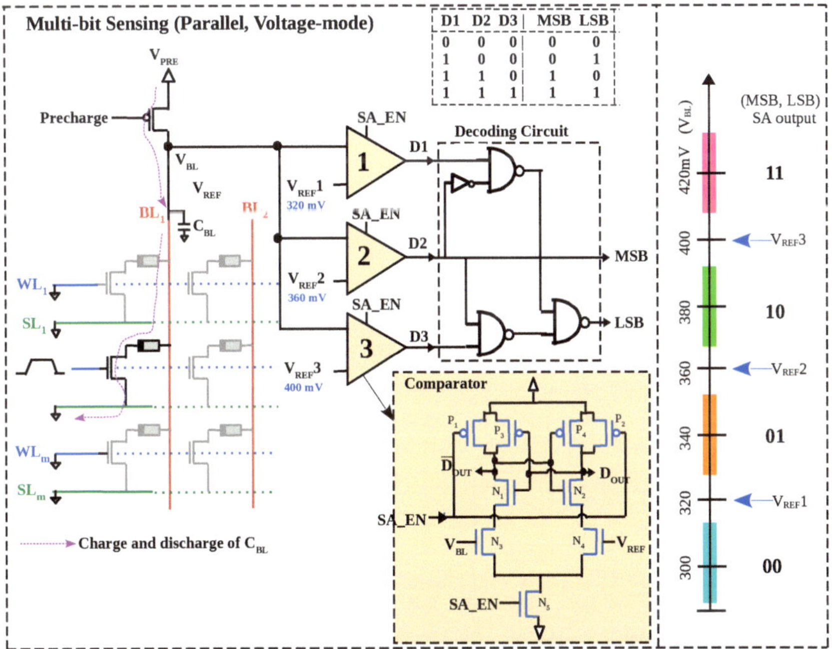

Fig. 7.7 In Parallel-sensing approach to multi-bit sensing, the cell's resistance is converted to a voltage and compared with multple V_{REF}s simultaneosuly. The output of the comparison is converted to (MSB,LSB) using a decoding logic

7.3.2.1 Sense Amplifier for multi-bit data

To illustrate sensing of multiple states, we consider the circuit shown in Fig. 7.7. This circuit belongs to the parallel-sensing type and the mode of sensing is voltage mode. The principle of operation is same as that for a single-bit sensing in voltage-mode. Hence we do not discuss it in detail and focus only on the V_{BL} to (MSB, LSB) conversion. As in single-bit sensing, the BL is pre-charged to a fixed voltage and discharged through the ReRAM cell. Depending on the state of the ReRAM, V_{BL} gets discharged from V_{PRE} to any of the four voltages (corresponding to four resistive states). For the purpose of illustration, let us assume that the V_{BL} gets discharged from V_{PRE} to 300 mV, 340 mV, 380 mV and 420 mV which should be sensed as (00),(01),(10) and (11), respectively. The V_{BL} is fed to three comparators. In parallel-sensing, to sense n-bit, we need (2^n-1) comparators. The comparators output $D1, D2, D3$ which is a '1' if $V_{BL} > V_{REF}$ and a '0' if $V_{BL} < V_{REF}$. In comparator 1,2 and 3, V_{BL} is compared with 320 mV, 360 mV and 400 mV, respectively (Fig. 7.7). If V_{BL} is 340 mV (or somewhere near it), $D1$ will be '1' and $D2$ and $D3$ will be '0'. The decoding circuit converts $D1, D2, D3$='100' to (MSB=0, LSB=1). In this manner, the BL voltages are converted to 2-bit data during the sensing phase.

Other similar approaches to parallel-sensing for multi-bit data can be found in [11, 12]. Some circuits for sensing multiple states in time domain can be found in [13, 14].

It must be observed that using ReRAM technology to store multiple states (instead of two states) comes at a cost - the peripheral circuitry. For writing, as illustrated in Fig. 7.5, a single gate voltage is not enough. Three different gate voltages are needed and the row decoder must be able to drive the WLs to these different voltages during the SET process. This will require a sophisticated row-decoding circuitry *e.g.* level-shifters have to be incorporated into the row decoder to be able to drive the WL to different voltages during MLC writing. In the same manner, distinguishing between multiple states complicates the SA design. As depicted in Fig.7.7, to sense 2-bit data, three V_{REF}s/comparators are needed and this will increase the area of the READ circuit. Consequently, the peripheral circuitry of a ReRAM array with MLC capability is a complicated circuit and this cost must be counted before using the MLC capability of this technology. Sample peripheral circuits for ReRAM array with MLC capability are presented in [15, 16].

7.4 Memory Architecture: A Brief Introduction

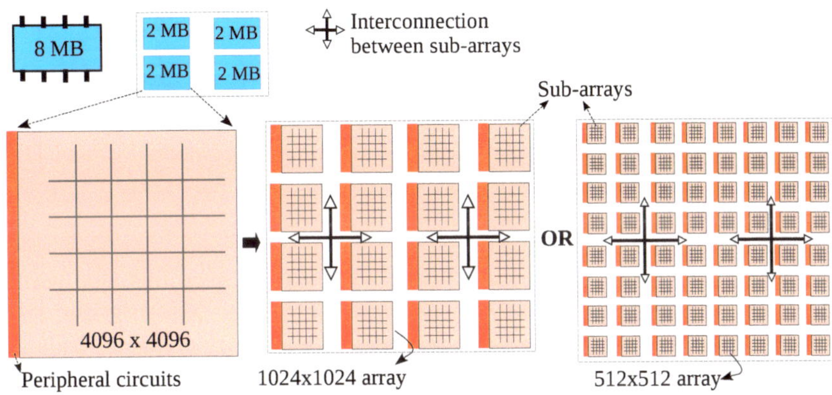

Fig. 7.8 A 4096×4096 array can be partitioned in many ways. Two ways are illustrated– sixteen 1024×1024 arrays or 64 arrays, each of size 512×512.

The memory architecture design is basically the art of partitioning the required memory size into many smaller memory arrays. Suppose you want to design a 8 Mega Byte (64 Mega Bit) memory using ReRAM. That would be 64×1024×1024 bits = 67,108,864 bits of information. If we design a symmetrical array (equal rows and columns), storing such a large quantity of bits will require a 8192×8192 array! That would be impractical (for reasons that follow). Let us say, we split 8 MB memory as four 2MB memory blocks. 2 MB = 16,777,216 bits. Storing 2 MB of bits will require a 4096×4096 array. Designing such a large array has the following problems:

1. The long wires of row and column will make the access time large due to wire parasitics. This will affect the speed of READ/WRITE operations since the processor will usually be faster than the memory.
2. The power consumption will be huge since the entire array has to switched ON while reading/writing into a small portion of the array.

In contrast, partitioning the required memory into smaller arrays (called sub-arrays) will keep the wire length within bounds, resulting in faster access times. Furthermore, all the arrays (with their peripheral circuits) need not be switched ON, but can be in power-saving mode. Typically, sense amplifiers, row/column decoders and WRITE circuits consume huge power and they can be disabled when the corresponding sub-array is not active [17]. This will enable the entire memory chip to have faster access and lower power consumption. Fig. 7.8 illustrates the partitioning of a 4096×4096 array. The long wires of the WL (row) and BL (column) of a 4096×4096 array are broken into smaller wire segments, thus reducing the wire length. Usually the long wire of WL and BL are split symmetrically along horizontal and vertical directions into equal number of sub-arrays along both directions (Fig. 7.8). This is because WL selection and pre-charging during sensing require charging the WL and BL capacitance, respectively. The obvious disadvantage of partitioning a larger array into smaller sub-arrays is the area. As illustrated in Fig. 7.8, separate peripheral circuits are needed for each sub-array, increasing the total area of the memory chip. Furthermore, there needs to be interconnection wires between sub-arrays, carrying synchronization signals. Nevertheless, the advantages outweigh the area disadvantage and memory chips are usually designed in a modular fashion – smaller sub-arrays grouped into a block and many such blocks forming a bigger block till the required memory capacity is achieved. Given a memory size (2 MB), should it be implemented as 64 sub-arrays of size 512×512 or 256 sub-arrays of size 256×256? This is a complex design problem and answering it is beyond the scope of this book. Memory architects depend on array and peripheral circuits models to conduct numerous simulation studies before finalizing the size of the sub-array. In the work in [18], researchers have observed (by simulations) that array access latency increases from 200-300 ns to 2 μs as the size of the sub-array increases from 64×64 to 1024×1024. Other related research in this direction include [19] where the effect of programming current on the size of the sub-array is studied. The power consumption as the size of the sub-array increases is studied in [20].

7.4.1 Sub-array Design

Fig. 7.9 depicts a 1k bit ReRAM array with its peripheral circuitry. In general, a sub-array is a small array with its own peripheral circuitry* (Row decoder, READ/WRITE circuits, BL/SL Mux) [22]. As depicted, an array of 16 rows and 64 columns forms a 1kbit sub-array. Usually, the

* There are exceptions to this general rule. To save area, adjacent sub-arrays can share a row decoder if they are not simultaneously accessed and this technique has been used in DRAM [1].

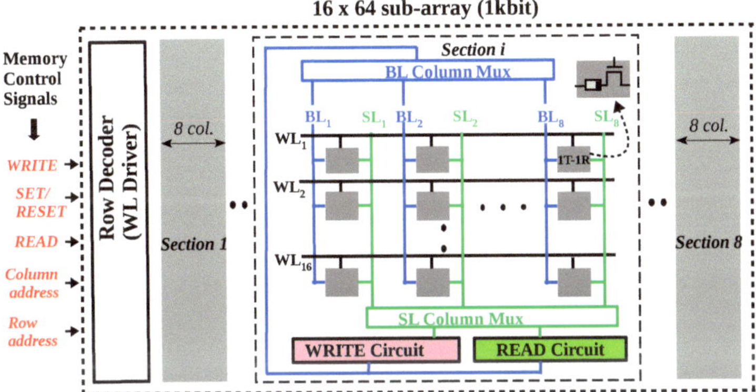

Fig. 7.9 A 16×64 sub-array is partitioned into sections (8 columns form a section) and each section has a dedicated READ and WRITE circuit [21]

WRITE circuits and READ circuits occupy a large area when compared to the memory array and it is difficult to have a dedicated READ and WRITE circuit for each column. This is true of 1T-1R and 1S-1R array configurations where the individual cell is of the order of $12F^2$ or $6F^2$. The WRITE circuit usually has an op-amp which occupies large area. Regarding the READ circuit, the area occupied depends on the type of the Sense Amplifier chosen. Nevertheless, even a compact SA will have at least 12 MOS transistors while the width of a single column corresponds to the width of a single transistor in 1T-1R configuration. Therefore READ and WRITE circuits are usually shared by columns [23, 24]. As depicted in Fig. 7.9, eight columns share a set of READ/WRITE circuits. The sub array is partitioned into sections (8 columns form a section) and each section has a dedicated READ and WRITE circuit. Each section has 16 rows and 8 columns of memory cells, having a memory capacity of 128 bits. Eight such sections have a memory capacity of 1kbit ($128 \times 8 = 1024$ bits). It must be noted that there is no standard way of designing sub-arrays and determining the optimal size of the section (within a sub-array) is a complex design problem similar to determining the optimal size of the sub-array. For example, if the READ and WRITE circuits are huge, it may be required that 16 columns share them, implying a section size of 16 columns. With a compact READ/WRITE circuit and a skilled layout engineer, it might be possible to have a READ/WRITE circuit for every 4 columns.

To conclude, a sneak peek of memory architecture was provided in this section. Memories (both standalone or embedded into a SoC) are designed in a modular fashion with sub-arrays being the indivisible array unit at the lowest level. Hence, each sub-array has a continuous arrangements of memory cells (rows×columns) with its own peripheral circuitry. Sub-arrays are grouped to form bigger blocks till the required memory capacity is achieved. At the moment, ReRAM chips are being developed in different architectures. Furthermore, with increasing trend towards in-memory computing (discussed in Part III), ReRAM is seldom designed as a pure memory but augmented with some computing capability.

References

[1] K. Itoh, "Dram circuits," in *VLSI Memory Chip Design*, pp. 97–194, Berlin, Heidelberg: Springer Berlin Heidelberg, 2001.

[2] A. Levisse, B. Giraud, J. P. Noel, M. Moreau, and J. M. Portal, "High density emerging resistive memories: What are the limits?," in *2017 IEEE 8th Latin American Symposium on Circuits & Systems (LASCAS)*, pp. 1–4, 2017.

[3] K. Abbas, "Memories," in *Handbook of Digital CMOS Technology, Circuits, and Systems*, pp. 471–518, Cham: Springer International Publishing, 2020.

[4] S. Yu, "Semiconductor memory technologies overview," in *Semiconductor Memory Devices and Circuits*, pp. 1–20, CRC Press, 2022.

[5] A. Prakash and H. Hwang, "Multilevel cell storage and resistance variability in resistive random access memory," *Physical Sciences Reviews*, vol. 1, no. 6, pp. –, 2016.

[6] Y. Luo, X. Han, Z. Ye, H. Barnaby, J.-S. Seo, and S. Yu, "Array-level programming of 3-bit per cell resistive memory and its application for deep neural network inference," *IEEE Transactions on Electron Devices*, vol. 67, no. 11, pp. 4621–4625, 2020.

[7] J. Reuben, D. Fey, and C. Wenger, "A modeling methodology for resistive ram based on stanford-pku model with extended multilevel capability," *IEEE Transactions on Nanotechnology*, vol. 18, pp. 647–656, 2019.

[8] C. Calligaro, V. Daniele, R. Gastaldi, A. Manstretta, and G. Torelli, "A new serial sensing approach for multistorage non-volatile memories," in *Records of the 1995 IEEE International Workshop on Memory Technology, Design and Testing*, pp. 21–26, 1995.

[9] C. Calligaro, R. Gastaldi, A. Manstretta, and G. Torelli, "A high-speed parallel sensing scheme for multi-level non-volatile memories," in *Proceedings. International Workshop on Memory Technology, Design and Testing (Cat. NO.97TB100159)*, pp. 96–101, 1997.

[10] M. Bauer, R. Alexis, G. Atwood, B. Baltar, A. Fazio, K. Frary, M. Hensel, M. Ishac, J. Javanifard, M. Landgraf, D. Leak, K. Loe, D. Mills, P. Ruby, R. Rozman, S. Sweha, S. Talreja, and K. Wojciechowski, "A multilevel-cell

32 mb flash memory," in *Proceedings ISSCC '95 - International Solid-State Circuits Conference*, pp. 132–133, 1995.

[11] Y. Yilmaz and P. Mazumder, "A drift-tolerant read/write scheme for multilevel memristor memory," *IEEE Transactions on Nanotechnology*, vol. 16, pp. 1016–1027, Nov 2017.

[12] F. Husain, B. Iqbal, and A. Grover, "A 0.4μa offset, 6ns sensing-time multi-level sense amplifier for resistive non-volatile memories in 65nm lstp technology," in *2021 34th International Conference on VLSI Design and 2021 20th International Conference on Embedded Systems (VLSID)*, pp. 76–81, 2021.

[13] J. Reuben and D. Fey, "A time-based sensing scheme for multi-level cell (mlc) resistive ram," in *2019 IEEE Nordic Circuits and Systems Conference (NORCAS): NORCHIP and International Symposium of System-on-Chip (SoC)*, pp. 1–6, 2019.

[14] X. Zhang, B.-K. An, and T. T.-H. Kim, "A robust time-based multi-level sensing circuit for resistive memory," *IEEE Transactions on Circuits and Systems I: Regular Papers*, vol. 70, no. 1, pp. 340–352, 2023.

[15] W. Li, X. Sun, S. Huang, H. Jiang, and S. Yu, "A 40-nm mlc-rram compute-in-memory macro with sparsity control, on-chip write-verify, and temperature-independent adc references," *IEEE Journal of Solid-State Circuits*, vol. 57, no. 9, pp. 2868–2877, 2022.

[16] W. Wan, R. Kubendran, C. Schaefer, S. B. Eryilmaz, W. Zhang, D. Wu, S. Deiss, P. Raina, H. Qian, B. Gao, S. Joshi, H. Wu, H.-S. P. Wong, and G. Cauwenberghs, "A compute-in-memory chip based on resistive random-access memory," *Nature*, vol. 608, no. 7923.

[17] J. M. Rabaey, *Digital integrated circuits: a design perspective.* USA: Prentice-Hall, Inc., 1996.

[18] M. Jagasivamani, C. Walden, D. Singh, L. Kang, S. Li, M. Asnaashari, S. Dubois, B. Jacob, and D. Yeung, "Memory-systems challenges in realizing monolithic computers," in *Proceedings of the International Symposium on Memory Systems*, MEMSYS '18, (New York, NY, USA), p. 98–104, Association for Computing Machinery, 2018.

[19] A. Levisse, p. Royer, B. Giraud, J. Noel, M. Moreau, and J. Portal, "Architecture, design and technology guidelines for crosspoint memories," in *2017 IEEE/ACM International Symposium on Nanoscale Architectures (NANOARCH)*, pp. 55–60, 2017.

[20] P. Narayanan, G. W. Burr, R. S. Shenoy, S. Stephens, K. Virwani, A. Padilla, B. N. Kurdi, and K. Gopalakrishnan, "Exploring the design space for crossbar arrays built with mixed-ionic-electronic-conduction (miec) access devices," *IEEE Journal of the Electron Devices Society*, vol. 3, no. 5, pp. 423–434, 2015.

[21] P. Chen, S. Sheu, K. Cheng, H. Lee, F. T. Chen, P. Chiang, M. Tsai, W. Lin, M. Chang, and Y. Chen, "Fast-write resistive ram (rram) for embedded applications," *IEEE Design & Test of Computers*, vol. 28, pp. 64–71, jan 2011.

[22] S. Yu, "Rram array architecture," in *Resistive Random Access Memory (RRAM): From Devices to Array Architectures*, pp. 35–54, Cham: Springer International Publishing, 2016.

[23] S. Yin, Y. Kim, X. Han, H. Barnaby, S. Yu, Y. Luo, W. He, X. Sun, J.-J. Kim, and J.-s. Seo, "Monolithically integrated rram- and cmos-based in-memory

computing optimizations for efficient deep learning," *IEEE Micro*, vol. 39, no. 6, pp. 54–63, 2019.

[24] J. Reuben and S. Pechmann, "Accelerated addition in resistive ram array using parallel-friendly majority gates," *IEEE Transactions on Very Large Scale Integration (VLSI) Systems*, vol. 29, no. 6, pp. 1108–1121, 2021.

Part III
IN–MEMORY COMPUTING

Chapter 8
In-Memory Computing: Applications of Resistive RAM Beyond Memory

8.1 Introduction

The reader might recall the statements made in the introduction of this book about the applications of Resistive RAM technology beyond a pure memory. Specifically, it was discussed how the conventional role of memory is being re-engineered in the field of integrated circuits. Historically, memory was considered a dumb device with no capability to compute or manipulate data which it stores. When you read from a memory, you get what you have written into it before. In other words, the sole purpose of memory (volatile or non-volatile) was storing or keeping the data unaltered in a chip. In DRAM, the charge on the capacitor is kept unaltered and in ReRAM, the resistance remains unaltered. As long as the device can maintain that state unaltered, it satisfied the requirements expected of a memory. Hence, the requirement for a memory was simply a storage unit and the conventional memories satisfied that requirement well, for many decades. Today, what is expected of a memory device in an integrated circuit is not just store data, but also perform some computations. This is due to a phenomenon called 'von Neumann bottleneck' in conventional computer architecture.

8.1.1 Von Neumann Bottleneck (The Memory Wall)

The movement of data between processing and memory units in present day computing systems is their main performance and energy-efficiency bottleneck, often referred to as the 'von Neumann bottleneck' or 'memory

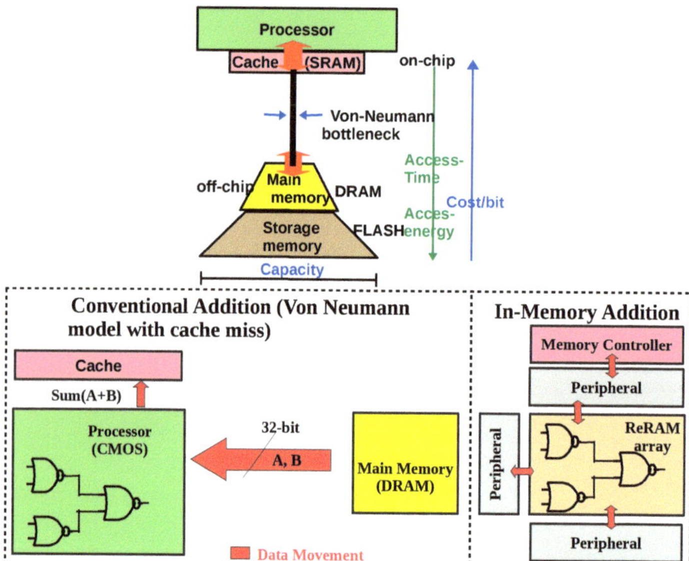

Fig. 8.1 (top) The von Neumann bottleneck. If the data required by the processor is not available in cache, it has to be brought from the main memory through a bandwidth-limited channel (bottom). If two 32-bit numbers have to be added, they have to be fetched from the DRAM and added in a CMOS-based processor in conventional addition. In in-memory addition, the data is moved only around the memory (array, peripheral circuitry and memory controller) and addition is performed in the array.

wall'. The memory wall problem has two facets: the mismatch in the performance (speed) of processor and memory and the energy for data transfer during memory access. The energy to access (transfer) data is growing exponentially along the memory hierarchy (from cache to off–chip DRAM, Fig.8.1-(top)). As a quantitative example, Stanford university researchers [1] point out that the energy for DRAM access is 3556 × the energy for 16-bit addition in 45 nm CMOS technology. It is widely accepted that 'data movement energy' dominates the 'computation energy' in traditional systems with separate memory and processing units, *i.e.* the computation in itself consumes only a small fraction of the energy [1, 2, 3] . There has been an ongoing effort (since 15-20 years) to combat the memory wall by bringing the processor and memory unit closer to each other. Early researchers used the term processing-in-memory (PIM) to refer to the effort to move the processing closer to where data resides. The term processing-in-memory broadly referred to processing in the memory using computing units placed in the memory chip. Later researchers used the term near-memory computing or near data processing to refer to the same effort and they exploited 3D stacking of DRAM dies over logic die to compute near memory. However, all these efforts try to minimize the physical distance between memory (where data is stored) and processing units and they do not completely solve the von Neumann bottleneck because the **need for data transfer persists**. The era of big data exacerbates the von Neumann bottleneck. Big data signifies increase in volume (due to increase in number of communication devices), variety (photos, videos, audio recordings, email messages, documents, books, presentations) and velocity of data (data traffic was expected to grow by a factor of 1455 from 2018 to 2020) [4]. This increase in data poses challenges to the conventional von Neumann architecture where memory and processor are separated.

The memory wall faced by computer architects necessitated a paradigm shift in the way data is processed. At the moment, there is an increasing trend to move computing to the location of data, a memory-centric computing [5, 6, 7]. The term 'in-memory computing' (also called 'processing-in-memory','compute-in-memory', 'logic-in-memory') is used to refer to any effort to process data at the residence of data (*i.e.* in the memory array) without moving them to a separate CMOS-based processing unit. If one can compute at the residence of data, in principle, the memory wall problem could be solved since the demarcation (between processing unit and memory) and the accompanying costly data transfer can be eliminated. The data transfer between processor and memory is costly in terms of energy and latency. Furthermore, when we move computing from processor to the memory, the substrate on which computing is performed is usually the memory array. The regular structure of the memory array (rows and columns arrangement of memory cells) is advantageous for parallel-processing and may even aid in reducing the latency of computation in certain cases, *i.e.* numerous operations can be performed simultaneously in the columns of a memory array which is not possible in typical CMOS substrate constructed from heterogeneous gates. In-memory computing is being explored for a wide variety of applications like scientific computing (linear and partial differential equations), machine learning, signal/image processing and pattern recognition. In all these applications, arithmetic (addition/subtraction) serves as the foundation. Detailed survey of in-memory computing using emerging NVMs are presented in [5, 7, 6, 8, 9]. In this chapter, we will limit our discussions to 'In-memory Arithmetic' and 'Matrix Vector Multiplication'.

8.2 In-memory Arithmetic

The author uses the term "arithmetic" as it is used in computer arithmetic. However, here the arithmetic is performed completely in the memory array. Addition, being the basic component of all computer arithmetic, this section will survey adders implemented in ReRAM array. Boolean logic has been the predominant way of designing adders in CMOS technology. Adders were synthesized in terms of logic gates, optimized using logic synthesis tools and mapped to 'x' nm node during technology mapping. Each gate is implemented as a pull-up (PMOS) and pull-down (NMOS) network of MOS transistors. Hence, the atomic computing unit in CMOS adders is the transistor (the latency and switching energy of transistors dictated the speed and power consumption of the adder). When researchers wanted to move arithmetic to memory, they followed a different path. They pursued a method to implement a Boolean logic gate using memory cell(s) of the array while minimally modifying the array and/or its peripherals. Adders could then be implemented as a sequence of Boolean logic operations in this array. This implies that the atomic computing unit of in-memory adders is the logic primitive or the Boolean gate (there may be exceptions). Hence, logic primitive plays a significant role in the performance of in-memory adders. In addition to logic primitive, the way in which carry is propagated (ripple carry, carry look-ahead, parallel-prefix *etc.*) also plays a crucial role in the performance of in-memory adders. The latency of different adders heavily depends on the method of carry-propagation.

8.2.1 Boolean Logic Gates in ReRAM array

As stated, adders are implemented in memory array by implementing Boolean logic primitive in the memory array. To implement complex functions like adders, the Boolean logic primitive chosen must be 'functionally complete'. A logic primitive or a set of logic primitives is said to be functionally complete if any Boolean function can be implemented in terms of this logic primitive (s). For example, NOR is 'functionally complete' since any Boolean function can be expressed in terms of NOR logic primitive*. Therefore, if a NOR gate can be implemented in a ReRAM array, any Boolean function can be implemented in the ReRAM array. We call this

* Recall from Digital Logic course that NAND and NOR gates were called universal gates for this reason.

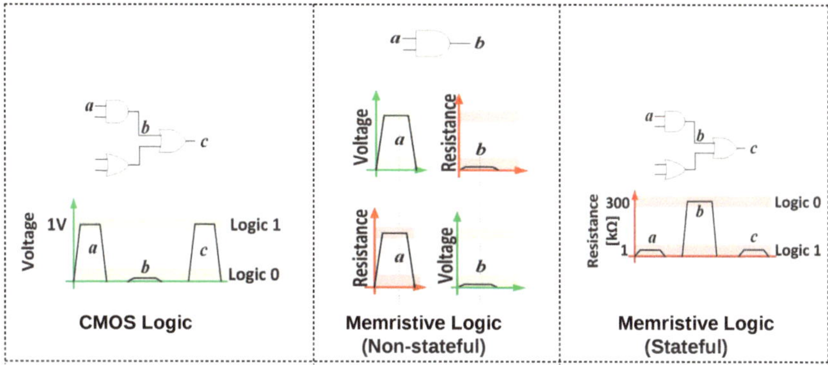

Fig. 8.2 In CMOS logic, inputs, outputs and intermediate values are represented as voltages; in memristive logic, inputs, outputs and intermediate values are represented as either voltages or resistances. A memristive logic is said to be stateful if resistance is the only state variable for representing inputs, outputs and intermediate results of computation.

type of logic 'memristive logic' since a memristive device (Resistive RAM) was used to construct it. Other than NOR, NAND, IMPLY+FALSE [10] and Majority+NOT [11] are also functionally complete. Before we discuss the implementation of some memristive logic gates, we must clarify an important difference between these memristive logic gates and CMOS logic gates. In CMOS logic, there is only a single logic state variable, *i.e.*, voltage. The input data is represented as voltage and is processed as voltage throughout the computation (including in all of the intermediate stages), and is finally also represented as voltage at the output, as illustrated in Fig.8.2. Furthermore, the state variable is regenerated throughout the computation by the CMOS gates. This seamless flow is disrupted in memristive logic gate because the internal state of memory cells governs their resistance, introducing resistance, in addition to voltage, as a logic state variable for computation [12]. When Boolean logic is implemented by manipulating the state (resistance) of the memristor, it is called stateful logic [13]. In a stateful logic family, both the input and the output of the computation is resistance. In certain logic families, the input is a voltage, but the output is a resistance [14] and in certain other families, the input is the resistance and the output of computation is a voltage [15]. These are classified as non-stateful logic families (Fig. 8.2). Due to space constraints, we discuss only one example for stateful logic and non-stateful logic here.

The reader is referred to [5, 8, 9] for other Boolean gate implementations in memory.

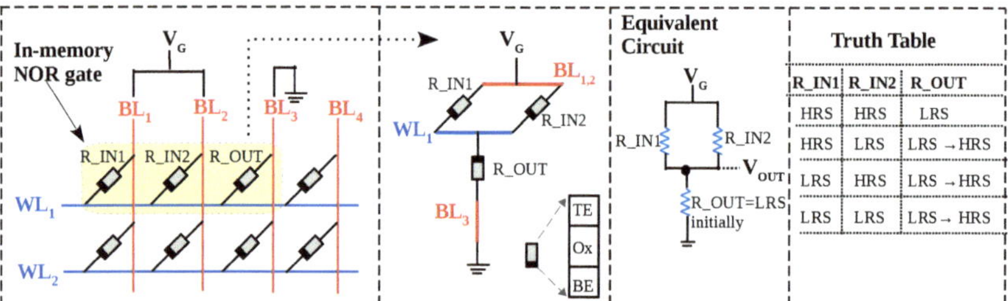

Fig. 8.3 NOR gate can be implemented by exploiting the fact that resistances R_{IN1} and R_{IN2} will be in parallel in the memory array. R_{OUT} which is initialized to LRS will either switch to HRS or remain in the same state, depending on inputs R_{IN1} and R_{IN2} of the NOR gate which are stored as resistance of the memory cell.

8.2.1.1 NOR Logic Operation in Memory Array

The implementation of NOR gate in the transistor-less memory array is depicted in Fig.8.3. It is called MAGIC (Memristor Aided LogIC) gate in literature [16]. This is a stateful logic gate meaning that both the input and output of the NOR gate (after completion of logic operation) are resistances of the memory cell. It is assumed that the inputs of the NOR gate are stored as resistances in two cells R_{IN1} and R_{IN2}. The final output of the NOR gate will be the resistance of R_{OUT} cell. The NOR gate can be implemented in two cycles. In the first cycle, the cell R_{OUT} is initialized to LRS. In the second cycle, a voltage V_G is applied to the top electrode of the two cells which hold the input resistances R_{IN1} and R_{IN2}. At the same time, the top electrode of R_{OUT} is grounded. As depicted in Fig. 8.3, this results in a topology where R_{IN1} and R_{IN2} form a voltage divider with R_{OUT}. Essentially, R_{IN1} is in parallel to R_{IN2}. When one of the input resistances is LRS, the equivalent resistance ($R_{IN1}||R_{IN2}$) is less than LRS and hence V_{OUT} will be approximately $V_G/2$. If V_G is chosen such that it is greater than $2.|V_{RESET}|$, this will result in the R_{OUT} switching from LRS to HRS (since $|V_{RESET}|$ will appear across R_{OUT}). For the case when both input resistances are HRS, the entire voltage will drop across the parallel combination of R_{IN1} and R_{IN2} (which is $\frac{HRS}{2}$) and V_{OUT} will be approximately 0 volts. This is because the ReRAM device is chosen such that its HRS is much greater than its LRS ($\frac{HRS}{LRS} \geq 100$).

The execution of aforementioned NOR gate needs specific ReRAM characteristics since it needs a device with large $\frac{HRS}{LRS}$ ratio and also a device whose $\frac{V_{SET}}{V_{RESET}} > 2$ [17, 6]. Also special isolation voltages are needed to be applied in rows/columns where NOR operation is not intended to lower sneak-path currents and avoid write-disturb phenomenon. The advantage of this in-memory NOR gate is parallelism, *i.e.* multiple NOR gates can be simultaneously executed in rows/columns of the memory array. Although single NOR gate is demonstrated in [17], successful implementation of executing many NOR gates in parallel is not reported. Perhaps the most serious disadvantage of this logic gate is that the NOR logic primitive is a weak primitive, *i.e.* compared to logic primitives like MAJ and XOR, NOR requires more gates to express the same Boolean function. As the complexity of the circuit increases, the latency increases since the complex function has to executed in the memory array as a series of NOR operations.

8.2.1.2 Majority Logic Operation in Memory Array

Majority logic, a type of Boolean logic, is defined to be true if more than half of the n inputs are true, where n is odd. Hence, a majority gate is a democratic gate and can be expressed in terms of Boolean AND/OR as $MAJ(a, b, c) = a \cdot b + b \cdot c + a \cdot c$, where a, b, c are Boolean variables. Although majority logic was known since 1960, there has been a revival in using it for computation in many emerging nanotechnologies (spin waves, magnetic Quantum-Dot cellular automata, nano magnetic logic, Single Electron Tunneling). Recent research [11, 18, 19, 20] has confirmed that majority logic is to be preferred not only because a particular nanotechnology can realize it, but also because of its ability to implement arithmetic-intensive circuits with less gates.

In [23, 21, 22], a majority logic gate is implemented in a 1T–1R array in a non-stateful manner, *i.e.*, the inputs of the majority gate are the resistances of the cells and the output is sensed as a voltage. Consider an array of ReRAM cells arranged in a 1T-1R configuration, as depicted in Figure 8.4. Each cell can be individually read/written into by activating the corresponding wordline (WL) and applying appropriate voltage across the cell (BL and SL). Now, if three rows are activated simultaneously during read operation (Rows 1 to 3 in Figure 8.4, the

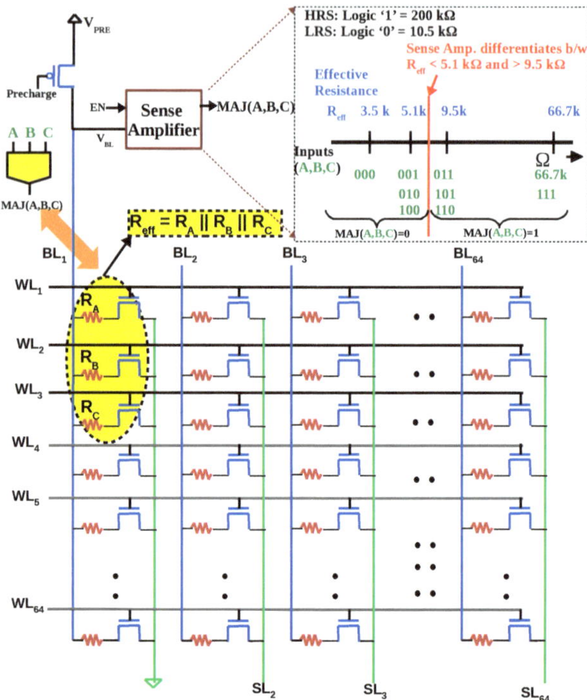

Fig. 8.4 In-memory implementation of majority gate [21, 22]: In a 1T-1R array, the resistances (R_A, R_B, R_C) in the three rows will be parallel if three rows are selected at the same time. (Inputs of the majority gate A, B, C are represented as resistances R_A, R_B, R_C). During READ, the effective resistance R_{eff} can be accurately sensed to implement an in-memory majority gate.

resistances in column 1 are in parallel (neglecting the parasitic resistance of BL and SL). The effective resistance between BL and SL will therefore be $R_{eff} = (R_A + r_{DS})||(R_B + r_{DS})||(R_C + r_{DS}) \approx (R_A||R_B||R_C)$, if the drain-to-source resistance of transistor (r_{DS}) is small compared to LRS. A Sense Amplifier (SA) which can accurately sense the effective resistance implements a 'in-memory' majority gate. Table 8.1 lists the truth table of a 3-input majority gate and the effective resistance for all the eight possibilities. If we assume a LRS and HRS of 10.5 kΩ and 200 kΩ, respectively ([24]), the crucial aspect of the proposed gate is to be able to differentiate between R_{eff}^{001} (two LRS and one HRS) and R_{eff}^{110} (two HRS and one LRS). In other words, resistance $\leq$ 5.1 kΩ must be sensed as '0' and resistance $\geq$ 9.5 kΩ must be sensed as '1'. If we call the resistance to be differentiated as sensing window (9.5 kΩ – 5.1 kΩ = 4.4 kΩ), any sense amplifier which can differentiate this sensing window can be used to implement the majority gate. A current-mode SA was used in [23] and a time-based SA was used in [22] to verify the correct functioning of majority gate, even in the presence of reasonable ReRAM variations.

Table 8.1 Precisely sensing R_{eff} results in majority: Logic '0' is LRS (10.5 kΩ) and logic '1' is HRS (200 kΩ). Sense amplifier distinguishes between $R_{eff} < 5.1$ kΩ and $R_{eff} > 9.5$ kΩ.

A	B	C	$MAJ(A,B,C)$	R_{eff}	R_{eff}
0	0	0	**0**	$\frac{LRS}{3}$	3.5 kΩ
0	0	1	**0**	$\frac{HRS \cdot LRS}{LRS+2 \cdot HRS}$	5.1 kΩ
0	1	0	**0**	$\frac{HRS \cdot LRS}{LRS+2 \cdot HRS}$	5.1 kΩ
0	1	1	**1**	$\frac{HRS \cdot LRS}{HRS+2 \cdot LRS}$	9.5 kΩ
1	0	0	**0**	$\frac{HRS \cdot LRS}{LRS+2 \cdot HRS}$	5.1 kΩ
1	0	1	**1**	$\frac{HRS \cdot LRS}{HRS+2 \cdot LRS}$	9.5 kΩ
1	1	0	**1**	$\frac{HRS \cdot LRS}{HRS+2 \cdot LRS}$	9.5 kΩ
1	1	1	**1**	$\frac{HRS}{3}$	66.7 kΩ

It must be noted unlike NAND and NOR, majority as a logic primitive is not functionally complete. However, it forms a functionally complete logic when used together with NOT, *i.e.*, any Boolean logic can be expressed in terms of majority and NOT gates [11]. A NOT gate can be implemented by latching the inverted output of the SA during a normal READ operation, as illustrated in Figure 8.5. Similarly, if $\overline{MAJ}$ is required, it can be obtained by choosing between the read-out data and its complement during majority operation. Any boolean logic can be implemented as a sequence of READ (majority) and WRITE operations, as illustrated in Fig. 8.5. This majority gate has many advantages. Compared to the NOR gate presented in previous section, there is no switching or writing of the memory cell during logic operation. The logic operation is performed as a READ operation, which is very less energy-consuming (compared to WRITE operation) and also endurance-friendly. The reader might recall that ReRAM has limited endurance and hence being able to achieve the majority logic operation without switching the ReRAM cell is very significant from endurance point of view. Note that there will be power dissipation in the Sense Amplifier during majority operation, but

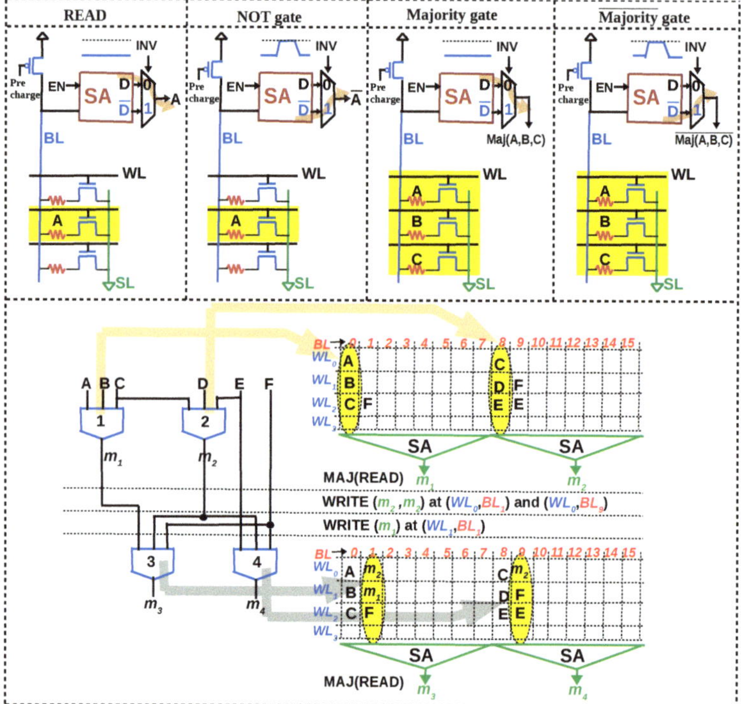

Fig. 8.5 NOT operation can be implemented by inverting the output of the SA during normal READ operation. Usually SAs based on strong Arm latch (Fig. 5.2, Fig. 5.3) output both the read-out data and its complement and hence NOT and $\overline{MAJ}$ can be obtained by simply choosing between the read-out data and its complement (top). By writing back the data, the array can be used to execute multiple levels of logic (bottom).

this energy consumption is only 0.6 pJ† while writing into a single ReRAM cell will cost 1 pJ -12 pJ [22]. Furthermore, this in-memory majority gate enables array-level parallelism, *i.e.* multiple majority gates can be executed simultaneously in the columns of the array (Fig. 8.5). Finally, the majority gate can be implemented in a 1T–1R array without necessitating any major change in the peripheral circuit (except the row decoder which needs to be modified to activate three rows simultaneously during logic operation).

† Note that this energy consumption is for a SA using a I_{READ} throughout sensing duration. As presented in Section 5.3, by pre-charging the BL and discharging it, this energy can be further reduced.

Note that this majority gate is non-stateful and similar non-stateful logic gates (OR,AND) have been demonstrated in 1T-1R arrays [25, 26].

8.2.2 In-memory Adders

Having discussed in-memory NOR and majority gates, we will discuss how complex arithmetic circuits can be implemented in memory. Since addition is the basic component of all computer arithmetic, this section will elaborate how a simple 1-bit full adder can be implemented in memory.

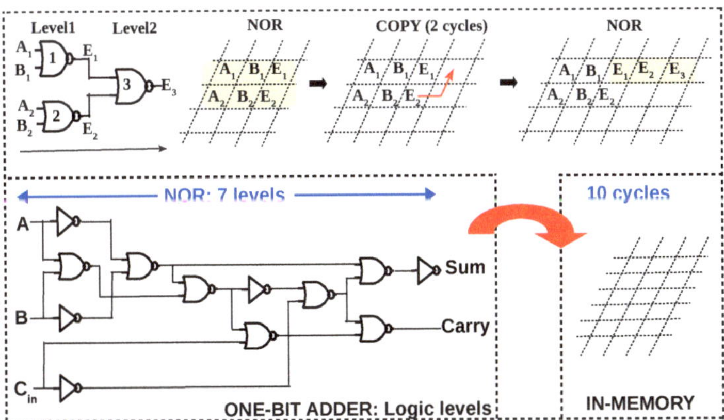

Fig. 8.6 Mapping a netlist of NOR gates and the COPY requirement due to the requirement that both the inputs of the NOR gate should be located in the same row. Due to this constraint while mapping to the memory, execution of the 1-bit full adder in the memory array requires 10 cycles.

8.2.2.1 One-bit Full Adder using In-memory NOR Gate

Although the in-memory NOR gate was simple to implement in two cycles (first initialise the output cell to LRS and then apply the gate voltage V_G for the execution of the NOR gate), the in-memory implementation of the full adder has its complications. The first obvious issue is the strict requirement of homogeneity, *i.e.* the entire circuit has to be synthesized solely in terms of NOR gates. The second issue is the difficulty in cascading in-memory gates. In conventional CMOS implementation, the inputs and outputs of a gate were voltages and cascading was seamless. Here, although the output of the first stage is resistance and the input of the next stage is also required as a resistance, it needs to be aligned with the input of the next gate. To illustrate, consider the two levels of NOR gates and the corresponding mapping to the memory array. As illustrated in Fig. 8.6, the cascading of two levels requires a COPY operation since the execution of NOR gate requires both the inputs to be in the same row - data locality constraint. COPY operation is usually accomplished as READ followed by a WRITE, requiring two cycles. In the initial experiments to map such circuits to the memory array, it was observed that 90% of the execution time was spent on data-arrangement. Therefore, the proponents of this NOR gate formulated a computer-aided algorithm to map any arbitrary circuit to the memory array [27, 28]. The algorithm takes a netlist of NOR gates and maps it optimally (in terms of latency and area) to the memory array while satisfying the data locality constraints and exploiting parallelism of gates in the input netlist. Using such a mapping algorithm [27, 28], the 1-bit full adder depicted in Fig. 8.6 was mapped to a 11×3 array and executed in 10 cycles [28].

8.2.2.2 One-bit Full Adder using In-memory Majority Gate

We recall that majority together with NOT forms a functionally complete logic *i.e.* any Boolean logic can be expressed in terms of majority and NOT gates [11]. Fig.8.7-(a) is a 1-bit full adder expressed in terms of majority and NOT gates [29]. Any Boolean logic can be synthesised in terms of majority and NOT gates using logic synthesis/optimization methods [11]. A 1-bit full adder can be executed in the 1T–1R array in six cycles, as elaborated in Fig.8.7-(b). It is assumed that the inputs to the full adder (A, B, C_{in}) are arranged in the memory array as depicted in the 4×16 array in Fig.8.7-(b). Further, it is assumed that 8 BLs share a SA, similar to a 1T–1R chip fabricated in [30]. The mapping from the circuit in Fig. 8.7-(a) to 1T–1R array is straightforward once the inputs to the gates are aligned in a column. Based on the input data dependencies of the gates, two gates can be executed in parallel in cycles 3 and 6 and the sum S and carry C_{out} bits will be available in the SA after 6 memory cycles. In cycle 3, $\overline{Majority(A, B, C_{in})}$ is needed at BL_8 because executing majority followed by NOT gate will increase the latency. Since the SA usually output the data and complement, $\overline{Majority}$ can be derived from $Majority$ without

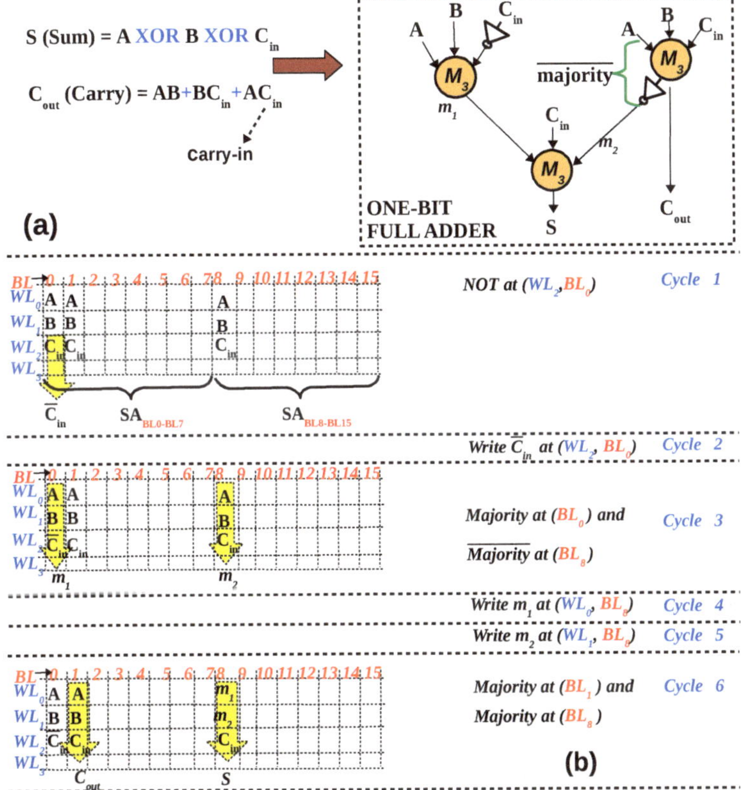

Fig. 8.7 (a) 1-bit full adder in majority logic, where M_3 represents 3-input majority gate [29] (b) Mapping of 1-bit adder to memory array. Both Majority and NOT are READ operations, as elaborated earlier in section 8.2.1.2

any extra cycle. When mapping large circuits, scheduling algorithms such as the ones used in high-level synthesis can be used to distribute the gates uniformly between levels such that gates at each level can be executed in parallel, thereby reducing latency.

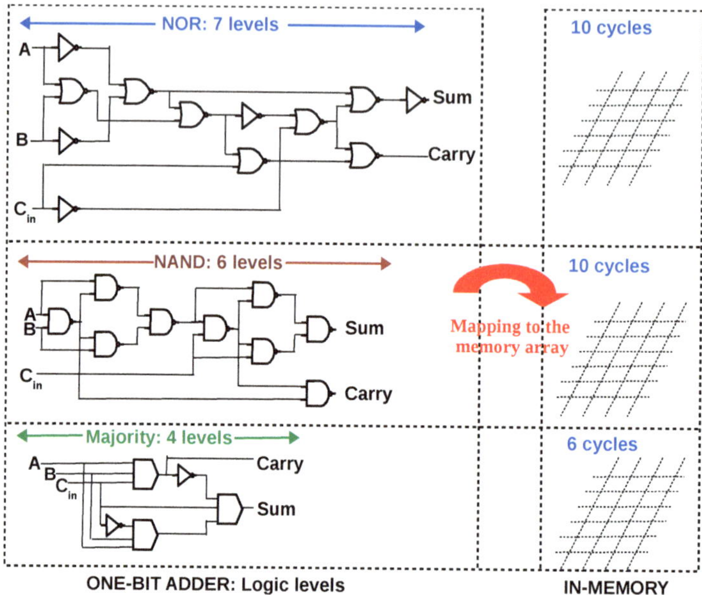

Fig. 8.8 n levels of Boolean logic will require $n + x$ cycles in-memory, where x depends on the characteristics of the memristive logic family. It must be noted that the number of cycles required (10 cycles for NOR, NAND and 6 cycles for MAJORITY) is already optimized by executing multiple gates in parallel (see the mapping for NOR, NAND and MAJORITY in ref [28],[31] and [23], respectively)

8.2.2.3 Optimizing Latency of In-memory Adders: Some Directions

The reader might have observed that, while implementing arithmetic circuits in memory, the latency of the mapped circuits is always more than the number of logic levels of the circuit before mapping. The seven levels of NOR gates required 10 cycles and the four levels of majority/NOT gates required 6 cycles for in-memory implementation. This increase in latency is due to the fact that cascading of gates in memory is not as straightforward as cascading of CMOS logic gates. Both in stateful and non-stateful logic families, this cascading inevitably increases the number of memory cycles needed to execute a function. A one-bit full adder in memristive logic family based on NOR [27], NAND [31] and MAJORITY [23] is compared in Fig. 8.8. It is evident that the number of steps (memory cycles) to compute in memory is larger than the number of logic levels.

When mapped to the memory array, n levels of logic will require $n + x$ cycles, where x depends on the characteristics of the memristive logic family. This includes attributes like statefulness, capability to execute gates in parallel *etc.* In a non-stateful logic family, the output of the gate may be a voltage and it may be needed as resistance for the input of the next level of logic, requiring an additional WRITE operation. In a stateful logic family, the output of the gate needs to be aligned with the inputs of following gate (next logic level), requiring an additional COPY/WRITE operation. In this manner, the interconnecting wires between logic levels contribute to additional cycles in memory. Furthermore, a memristive logic family should have the capability to execute multiple gates simultaneously. Consequently, multiple gates in a logic level can be mapped to the memory array in a single cycle. If the memristive logic family does not support the simultaneous execution of multiple gates, x will increase. Thus the parallel-friendliness of the logic family is an important characteristic to minimize latency.

Table 8.2 Latency comparison of in-memory adders (8-bit and n-bit)

Logic Primitive	Architecture	Latency	Latency (n-bit)	Ref.
IMPLY	Ripple carry	58	$5n + 18$	[10]
IMPLY	Parallel-serial	56	$5n + 16$	[32]
IMPLY + OR	Ripple carry	54	$6n + 6$	[33]
IMPLY	Semi-parallel	136	$17n$	[34]
NOR	Ripple carry	83	$10n + 3$	[35]
NOR	Look-Ahead	48	$5n + 8$	[36]
OR + AND	Ripple carry	49	$6n + 1$	[37]
ORNOR	Parallel-clocking	31	$2n + 15$	[38]
RIMP/NIMP	Pre-calculation	20	$2n + 4$	[39]
XOR	Ripple carry	18	$2n + 2$	[40]
XOR + MAJ	Ripple carry	18	$2n + 2$	[41]
XNOR/XOR	Carry-Select	9	–	[42]
OR + AND	Parallel-prefix	37	$8log_2(n) + 13$	[43]
Majority + NOT	Parallel-prefix	18	$4log_2(n) + 6$	[22]

To evaluate the efficiency of logic primitive used for in-memory computing, Table 8.2 analyses different in-memory adders that have been reported recently and their latency for 8-bit and n-bit. The adders are also classified based on the logic primitive and adder architecture. A key observation is that logic primitive plays an important role in determining the latency. IMPLY is a weak logic primitive, and generally incurs more latency than all other logic primitives. XOR and the majority are generally stronger logic primitives than OR/AND/NOR. This is evident from the fact that XOR-based and majority-based ripple-carry adders are faster than NAND/NOR-based ripple-carry adders [44]. In other words, with the adder configuration being the same (ripple-carry), logic primitive plays an important role in determining the latency of in-memory addition. Another important finding is that the adder architecture plays a key role in deciding the latency for increasing bit-width. This is evident from the latency of OR + AND-based adders in Table 8.2. Both adders used the same logic primitive (OR + AND), but [37] uses ripple-carry architecture, achieving a latency of $6n + 1$ while [43] uses a parallel-prefix configuration to achieve a latency of $8log_2(n) + 13$. As an example, a 32-bit adder based on OR + AND logic will require 193 cycles and 53 cycles for ripple-carry and parallel-prefix architectures, respectively. Hence, for larger bit-widths, architecture (carry propagation technique) plays an important role in latency. In summary, both adder architecture and logic primitive influence the latency of in-memory adder. Therefore, majority logic primitive and parallel-prefix adder architecture chosen in [22] drastically minimizes the latency of in-memory adders. The comparison of latency of adders listed in Table 8.2 is also plotted graphically in Fig. 8.9 to better visualize the increase in latency as the adder bit-width increases.

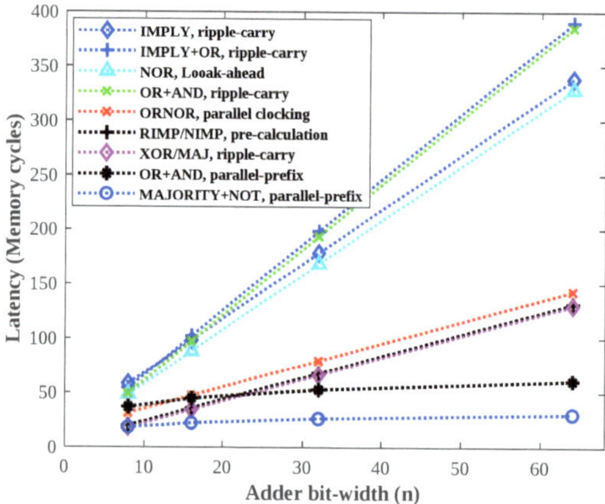

Fig. 8.9 Latency of in-memory adders with increasing bit-width, n. An adder with $O(log(n))$ latency is required for 32-bit/64-bit addition to harness the power of in-memory computation (Redrawn from [22]).

8.2.3 Eight-bit Parallel-prefix Adder in ReRAM array

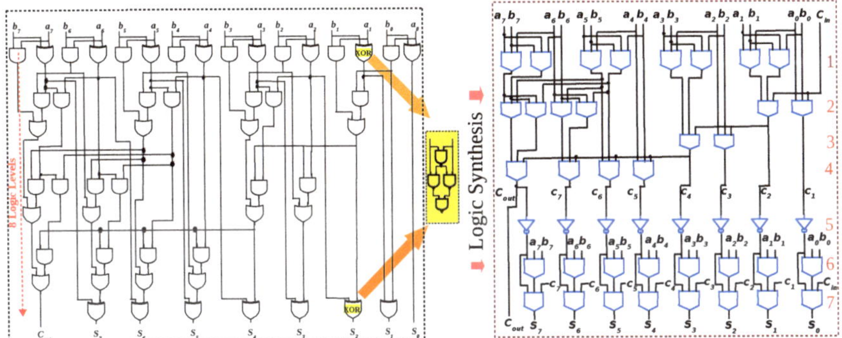

Fig. 8.10 Eight-bit PP adder of Ladner–Fischer type expressed in terms of AND, OR , XOR gates (left). Re-synthesized and optimized in terms of MAJORITY and NOT gates (right) [19, 20].

In this section, we discuss how an eight-bit adder is implemented using majority logic primitive and parallel-prefix architecture in the memory array. Conventionally, PP adders are synthesized in terms of AND, OR and XOR gates for CMOS implementation. Figure 8.10 (left) depicts an eight-bit PP adder of the Ladner–Fischer type. Three different logic primitives are required—AND, OR and XOR. As stated, different logic primitives require different modifications to the memory array and its peripheral circuitry. If a particular Boolean logic gate cannot be implemented in the memory array, it has to be re-formulated in terms of a logic gate that can be implemented in the memory array. For example, in the NAND-based logic family reported in [31], the XOR gate cannot be implemented; therefore, it is expressed as four NAND gates. As depicted in Figure 8.10 (left), a single XOR becomes three levels of NAND logic, increasing its latency. In contrast, by expressing a PP adder purely in terms of MAJORITY+NOT gates, the PP adder can be efficiently implemented in the memory array. Furthermore, a majority-based PP adder achieves a marginal reduction in logical depth compared to conventional AND-OR-XOR implementation (Figure 8.10). This is due to the majority being a stronger logic primitive than NAND/NOR/IMPLY [44]. To synthesize PP adders in terms of majority gates, logic synthesis tools can be used. A logic synthesis tool is proposed in [11], which takes any AND-OR-INVERT-based logic and synthesizes it purely in terms of majority and NOT gates. Boolean logic minimization techniques such as re-shaping, push-up, node merging, etc., are used to re-synthesize and optimize conventional AND-OR-INVERT logic in terms of MAJORITY-INVERT [45, 46, 47, 48, 49, 50]. Since the majority is the fundamental logic primitive for many emerging nanotechnologies, there are also works which pioneered the synthesis of PP adders solely in terms of majority gates. The reader is referred to [19, 20, 51] for such works. Therefore, a variety of techniques can be used to transform PP adders in terms of majority and NOT gates. Figure 8.10 (right) depicts a 8-bit PP adder, synthesized solely in terms of majority and NOT gates. In addition to achieving homogeneity, the majority-based PP adder incurs one level of reduction in logical depth compared to the AND-OR-XOR-based PP adder.

8.2.3.1 Mapping Methodology

Having synthesized the PP adder in terms of majority and NOT gates, they can be implemented in memory using the in-memory majority gate described in Section 8.2.1.2. The design of in-memory PP adders is generic and can be used to implement any PP adder [52]. However, in this section, the Ladner–Fischer adder of Figure 8.10 is chosen and the in-memory implementation (mapping) steps are elaborated. The mapping of the majority-based PP adder to the memory array can be treated as an optimization problem. Any optimization problem has objectives or goals, which should be achieved in the presence of certain constraints. The objectives of in-memory mapping are as follows:

1. Latency of in-memory PP adder must be minimized (O_1);
2. Energy consumption during addition must be minimized (O_2);
3. Area of the array used during computation must be minimized (O_3).

The aforementioned objectives are no different from the objectives of any VLSI circuit. All objectives cannot be met simultaneously in this mapping, and trade-offs must be made between latency of addition (O_1) and the area of array that is used (O_3). Any arithmetic circuit implemented in memory is bound to be very slow due to the high latency of in-memory adders. The latency of in-memory adders reported in the literature grows as $O(n)$ and 32-bit/64-bit adders in memory require hundreds of cycles [52]. Therefore, in this mapping, we focus on O_1 and minimize the latency. Minimizing the latency might result in the array area being compromised. However, latency is the more serious issue compared to array area in in-memory addition, for the following reasons:

1. A conventional adder (in CMOS) is devoted to addition while we re-use the existing memory array in in-memory computation. Hence, the increased array area required during addition is not a disadvantage, as long as computation can be performed in the memory array without an extra array;
2. ReRAM cell is a nano-device and does not significantly contribute to area. For example, a single 1T-1R cell with 130 nm MOS transistor occupies 0.2 μm^2 [53].

The constraints are specific to this logic family and can be summarized as follows:

1. Majority operation must be executed at three consecutive rows (C_1);
2. Due to the bounded endurance of ReRAM devices, the number of times a cell is switched must be minimized. (C_2).

C_1 must be satisfied during mapping because, during majority operation, three rows must simultaneously be selected. In principle, the three selected rows need not be contiguous and can be in different locations in the memory array (e.g., row 5, 8, 15 of a 64×64 array). However, row-decoding will become complicated. For practical in-memory implementation, the mapping must be 'peripheral circuit friendly'. In [22], a triple-row decoder is proposed for triple row-activation during majority operation. To implement this decoder, multiple single-row decoders were interleaved. Furthermore, the same row-decoder must be able to perform single-row decoding and triple-row decoding. This is because, during normal memory operation, a single row must be selected and, during majority, three rows must be selected. To this end, an address translator circuit is used in the row decoder, which seamlessly switches between single-row activation and triple-row activation. The triple-row decoder [22] is designed in such a way that only three consecutive rows can be selected. Therefore, while mapping, the inputs of the majority gate (to be executed in memory in the next step) must be written in three consecutive rows.

Constraint C_2 is posed by the ability of the memory device to endure the repeated switching of its state *i.e.* endurance. Experimentally reported endurance vary from 10^6 to 10^{12} for ReRAMs. Due to this limited endurance, the number of times a memory cell is switched during addition must be minimized. Having identified the objectives and constraints, we formulate a generic methodology to map any PP adder to the memory array. As stated, if the PP adder is available in terms of AND-OR-XOR gates, they must be re-synthesized in terms of the majority and NOT gates using logic synthesis techniques/tools. Given a majority-based PP adder, optimal in-memory implementation is an optimization problem—minimize O_1, while meeting C_1 and C_2.

The following steps implement the PP adder in the memory:

1. Start with Logic level 1;
2. Simultaneously execute all majority gates of a logic level in the columns of the array (O_1);
3. Write the outputs of the majority gates to the precise locations where they are needed in the next logic level such that all the majority gates of the following level can be executed simultaneously (O_1);
4. During Step 3, write the outputs of the majority gates to a new location and do not overwrite the existing data (C_2);
5. During Step 3, write the outputs of the majority gates to contiguous locations in the memory array (C_1);
6. Repeat Steps 2–5 for the remaining logic levels.

Figure 8.11 illustrates the mapping of an 8-bit PP adder to the memory array. Majority gates 1–8 of the first logic level are executed simultaneously in one memory cycle. Since we know that, at the next level, majority gates 9, 10, 11, 12, 13, 14 need to be executed, we write the outputs of the first logic level $(m_1, m_2, m_3, m_4, m_5, m_6, m_7, m_8)$ to the exact location where they will be needed. When we write the output of the majority gates back to the array, they are written in consecutive rows (C_1) and are not overwritten on existing data (C_2). The in-memory steps are highlighted in yellow in Figure 8.11. The in-memory steps corresponding to logic levels 1 and 2 are:

1. Majority at col. (1, 9, 26, 33, 42, 49, 58, 65) rows 4–6 as a READ operation;
2. Write $(m_1 m_1 m_3 m_5 m_7)$ at col. (2, 10, 34, 50, 59) , row 4;
3. Write $(m_2 m_2 m_4 m_6 m_8)$ at col. (2, 10, 34, 50, 59), row 5;
4. Write $(m_3 m_4 m_3 m_4)$ at col. (2, 10, 17, 25) , row 6;
5. Majority at col. (2, 10, 17, 25, 59, 73) rows 4–6 as a READ operation.
6. ....

In this manner, the seven logic levels of an 8-bit adder can be executed in memory in 18 cycles. Reader is referred to [22] for the complete mapping.

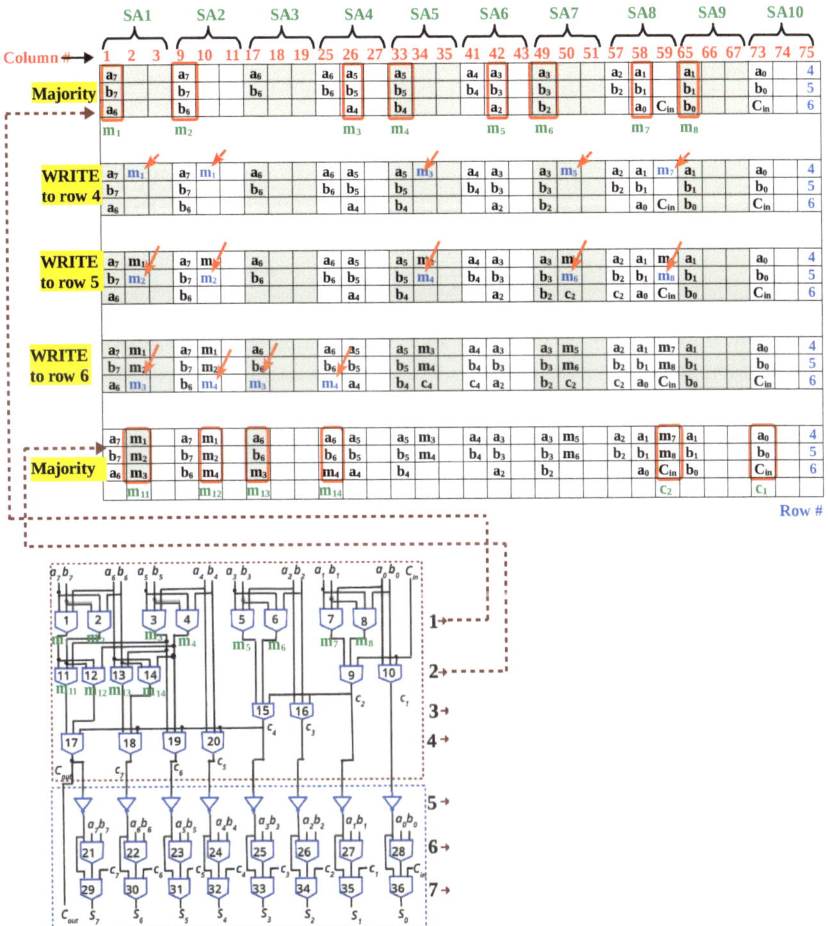

Fig. 8.11 Illustration of mapping of the first two logic levels to a memory array. Since each majority operation is executed as a READ operation, it can be written to the exact location it is needed at the next logic level while satisfying C_1 and C_2. In the above mapping, eight columns share a sense amplifier.

8.3 Matrix Vector Multiplication(MVM) in ReRAM Array

Perhaps the most sought-after application of Resistive RAM technology beyond memory is for performing MVM or MAC operations. This is because of the energy and latency saving in-memory implementation offers when compared to traditional CMOS implementation [54]. MVM/MAC is the fundamental and frequently required operation in neural networks, image processing, comibinatorial optimization, solving linear equations, sparse coding, associative memories, reservoir computing and many signal processing applications [6, 54]. According to [55], MVM/MAC operations can be as high as up to 90 % of the workloads for inference of Deep Neural Networks (DNNs).

Multiplication was performed in CMOS in a purely digital manner, *i.e.* the two numbers a and b to be multiplied were expressed in binary and the result was also in binary. Traditionally, multiplication was performed as AND operations leading to partial products, followed by summation of the partial products. Generally, multipliers are area intensive and latency intensive in CMOS technology. This is why pipelining and numerous architectures like Booth multiplier and Wallace tree multiplier were developed to optimise the partial-product generation and summation stages of a binary multiplier. The precision of multiplication is defined by the digital representation of the real numbers (number of bits, floating/fixed point) [54].

In contrast to such a digital implementation, multiplication was pursued in an analog manner in Resistive RAM technology[‡]. The analog conductance capability of ReRAM can be exploited for MVM. Furthermore, the inherent matrix-like structure of a memory array enables the straightforward mapping of the matrix entries as the conductance of the memory cells [56]. We consider a simple example for analog implementation of matrix multiplication in ReRAM array. More specifically, we take a basic neural network structure and illustrate how the computation, as illustrated in Fig. 8.12 can be implemented in the memory array.

In CMOS implementation, this would need three multipliers and three accumulators. An accumulator is nothing but an adder to accumulate the result, *i.e.* after calculating $x_1 \cdot W_{11}$, the result is stored in the accumulator and added to the next product $x_2 \cdot W_{21}$. If computing $x_1 \cdot W_{11}$ takes one multiplication cycle, then this MVM will take four multiplication cycles and three addition cycles in CMOS implementation (assuming three multipliers and three accumulators are available). It must noted that each multiplication cycle will in turn consume several sub-cycles since multiplication is a series of additions with carry propagation in between (marked by red arrows in Fig. 8.12). In contrast, the MVM implementation

[‡] Since research in MVM in ReRAM array is evolving, different implementations other than a purely analog implementation do exist.

in ReRAM array can be performed in just two steps. First, the weight matrix is written to the ReRAM as resistance of the individual cells. In the second step, the input vector is applied as voltages and the current in each column is summed and converted to an equivalent value using an ADC to get the digital result vector.

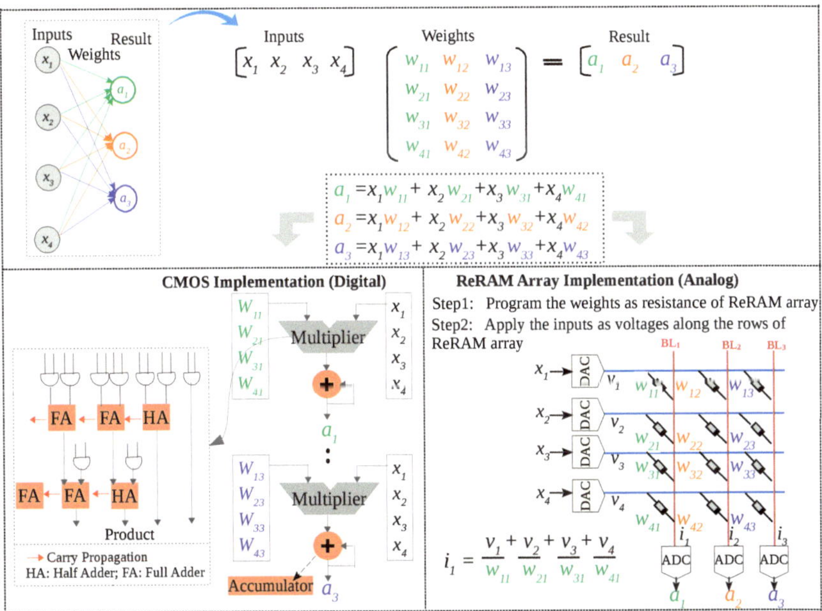

Fig. 8.12 A simple neural network and its computation as a Matrix Vector Multiplication (MVM). Conventionally, MVM is performed in CMOS as a series of multiplications and additions in a digital manner. In ReRAM array, the weights of the matrix can be programmed as resistances of individual cells and MVM can be implemented as READ operation of memory cells. When the input vector X is applied as a voltage at the rows of the memory array, each cell conducts a current in proportion to its resistance and the currents are summed up along the column to get the sum of products [57, 58, 54].

For sake of simplicity, let us assume the elements of the input vector X and the 4×3 weight vector W to be 2-bit integers. Let X be transformed as follows:

$$X_{ReRAM} = (X \cdot 0.1) + 0.1 \quad Volts \tag{8.1}$$

Then, in ReRAM array, X=[1 1 2 3] becomes [0.2 0.2 0.3 0.4] volts. To achieve correct MVM result, a higher weight must produce a larger current. Therefore, the weights must be mapped in an inverse order *i.e.* higher weights must be mapped to lower resistance. So, we map integers 0,1,2,3 to 800kΩ, 600kΩ, 400kΩ, 200kΩ, respectively. Thus X_{ReRAM} is X expressed in voltage domain and W_{ReRAM} is W expressed in resistance domain, as depicted in Fig. 8.13. The mapping of W_{ReRAM} to the rows and columns of the ReRAM array is non-trivial. The individual cells must be programmed to the required resistance(weight) by Incremental Program and Verify (IPV) algorithms discussed in Chapter 6. As stated in Chapter 6, the ReRAM cell can be programmed to different resistance during the SET process by varying the gate voltage of the access transistor (which varies the compliance current). This method must be used along with IPV algorithms to precisely program the ReRAM cell to the required resistance. Once the weight matrix is programmed as resistance of the cells, the array is ready for MVM operation.

As depicted in Fig. 8.13, X is applied as voltage along the rows. Each cell conducts current in proportion to its weight -higher the weight, larger the current through it. The currents are naturally added (due to Kirchoff's law) and the total current $[i_1 \quad i_2 \quad i_3]$ in each column is proportional to MVM result $[a_1 \quad a_2 \quad a_3]$. Since the ADC needs a voltage as its input, the BL current needs to be converted to an equivalent voltage using an I-to-V converter. BLs are connected to transimpedance amplifiers (TIA) to convert the output current of each column into voltage. The TIA additionally serves to hold the BL at virtual ground. Since BL_1, BL_2, BL_3 are at virtual ground, the current flowing in each cell is simply V_{WL}/R_{cell}, where $V_{WL} \in (0.1, 0.2, 0.3, 0.4)V$ and $R_{cell} \in (800k, 600k, 400k, 200k)\Omega$. For the example considered in Fig. 8.13, the reader can verify that

$$i_1 = \frac{0.2V}{600k\Omega} + \frac{0.2V}{400k\Omega} + \frac{0.3V}{600k\Omega} + \frac{0.4V}{600k\Omega} = 1.82 \quad \mu A$$
$$i_2 = \frac{0.2V}{400k\Omega} + \frac{0.2V}{400k\Omega} + \frac{0.3V}{200k\Omega} + \frac{0.4V}{400k\Omega} = 3.5 \quad \mu A$$
$$i_3 = \frac{0.2V}{600k\Omega} + \frac{0.2V}{400k\Omega} + \frac{0.3V}{400k\Omega} + \frac{0.4V}{200k\Omega} = 3.58 \quad \mu A$$

The reader can observe that i_1 is approximately half of i_2 and i_3. Thus we can expect that, after I-to-V conversion, with a good ADC, i_1 can be converted to 8 and i_2 and i_3 can be converted to 16. Similarly, $i_2 \approx i_3$ and an ADC can convert currents 3.5-3.6 μA to 16. In this manner, with an appropriate I-to-V converter (transimpedance amplifies in this case) and ADCs connected to the BLs, MVM can be achieved in a ReRAM array.

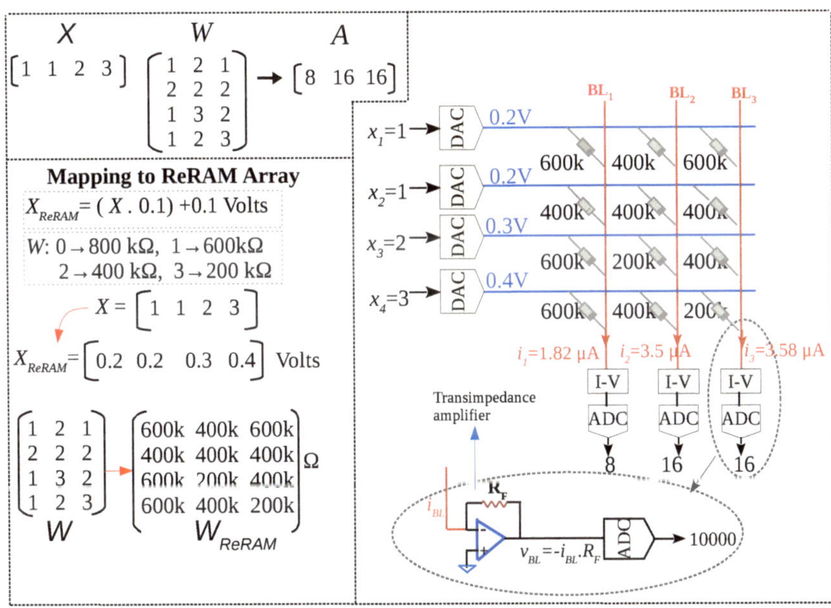

Fig. 8.13 Illustration of a simple MVM mapped to the memory array. An operational amplifier with a negative feedback resistor (R_F) is used as a transimpedance amplifier to bias the BL at 0 V and produce an output voltage, V_{BL} = -$i_{BL} \cdot R_F$, where i_{BL} is the current flowing out of BL [59],[60].

It must be noted that the peripheral circuitry of a ReRAM array augmented with the capability to perform MVM is much complicated when compared to the peripheral circuits of a normal memory array (presented in Part-II of this book). For MVM, the row decoder must not only support selection of multiple rows but also be able to drive the WLs to different voltages. Similarly, the read-out circuitry connected to the columns of a regular array was a simple Sense Amplifier (1-bit ADC) to convert the resistance to a bit. For a memory array to be capable of MVM, we need a transimpedance amplifier followed by an n-bit ADC for each column that needs to be simultaneously computed (columns of the matrix). The size of the ADC depends on the number of rows of the weight matrix and the number of bits used to encode the weight and the input. In this example, 2-bit X is multiplied with 2-bit W resulting in 4-bit output.

Since the matrix has four rows, four such 4-bit numbers have to be added resulting in a 6-bit output. The current at BL is maximum when X=(3 3 3 3) is multiplied with W^T=(3 3 3 3). This corresponds to $X.W^T = 36$ and the corresponding BL current is $4\times\frac{0.4V}{200k\Omega} = 8\mu$A. Similarly, the current at BL is minimum when X=(0 0 0 0) is multiplied with W^T=(0 0 0 0). This corresponds to $X.W^T = 0$ and the corresponding BL current is $4\times\frac{0.1V}{800k\Omega}$ $= 0.5\mu$A. Therefore the I-V converter and 6-bit ADC must be designed to convert currents in the range 0.5 μA–8 μA to (000000)–(100100). In general, for K-bit input and M-bit weight, MVM of size $N \times N$ requires an ADC of precision $\lceil log_2((2^K - 1) * (2^M - 1) * N)\rceil$ [59].

The advantages of in-memory implementation (in the ReRAM array) are as follows:
1. Only a single step is required for the multiplication ($I = V \times G$, where G is the conductance of the ReRAM) whereas the binary multiplication required sequence of MULTIPLY and ADD operations.
2. Energy consumption will be lower provided the peripheral circuits (ADC/DAC) are designed in an energy-efficient manner.
3. Finally, the movement of the inputs of the MVM/MAC operation (matrices) can be very energy consuming in case the matrices are stored off-chip (*e.g.* DRAM or tertiary storage made of NVM technology). This movement and the consequent energy consumption can be avoided if the input matrices are also stored in a neighboring memory array.

However, pioneering research in this field for the last decade has also identified some key disadvantages:
1. Interconnect resistance of the memory array (parasitic resistance of the WL and BL) degrade the accuracy of computation [60]. In addition to interconnect resistance, the non-linearity of the access device (selector/transistor) also affects the accuracy of the analog result.
2. Peripheral circuits (ADC/DAC) occupy large area when compared to the area occupied by the memory array. Hence, the peripheral circuit cannot be laid out in a pitch-matched manner *i.e.* the area occupied by an ADC will be much larger than the area occupied by a row/column. ADC/DACs are are also energy-consuming. It is estimated that ADCs consume upto 80% of the total energy and 70% of the area of a MVM core based on NVM crossbars [54, 61]
3. Precisely writing the weights as resistance of array is a challenge in most NVM technologies, especially in Resistive RAM. As stated, Incremental Program Verify algorithms are needed which consume energy and also take time. This is because, writing using IPV algorithms is not accomplished in a single pulse, but a series of pulses with READ pulses in-between.

Due to the ReRAM cell's non-idealities and other aforementioned effects, the accuracy of pure analog MVM is low. To overcome some of these disadvantages, MVM has also been explored using an old technique in computer architecture called 'Bit-slicing'.

8.3.1 MVM by Bit Slicing

Storing the weights of the matrix as a n-bit weight in the ReRAM array is a challenge. To store a n-bit weight, we need to program the memory cell to 2^n different resistive states. Understandably, for a 5-bit weight, we need to program the ReRAM cell to 32 different states which is impractical in the present technology due to variability and the overlapping of the resistance distribution between the neighboring states (typically in a ReRAM, as we accommodate more states, the distance between the states reduces, a constraint imposed by the physical switching mechanism). An alternative approach is to store the weights in binary form (each ReRAM cell being either HRS or LRS) and multiplying the output of each column by appropriate value outside the array. In essence, we are 'slicing' the computation over different columns of the array. This is similar to 'Bit-Slicing' in computer architecture in which a large computation is split into smaller modules where each module computes a slice of the computation, in parallel. Using Bit-slicing, MVM can be implemented in memory without programming the cell to multiple states and is also known to improve accuracy of computation [6]. The price one has to pay is the increased latency and the additional shift-and-add circuitry needed outside the memory array.

The process of multiplying input vector X with weight matrix W involves the following steps (the step number corresponds to the orange circled number in Fig.8.15):

1. Slice the weights into bits and program the memory array according to the weights *e.g.* a 3-bit weight 5 is sliced into bits '101' and written to the memory as (10 kΩ 200 kΩ 10 kΩ).
2. Slice the inputs into bits and apply them along the rows, bit by bit. Least significant bit(LSB) of all the elements of X are applied to the rows simultaneously.
3. Shift and add the outputs of the column, according to their position and store them in a register. This is Partial Product (PP_0).

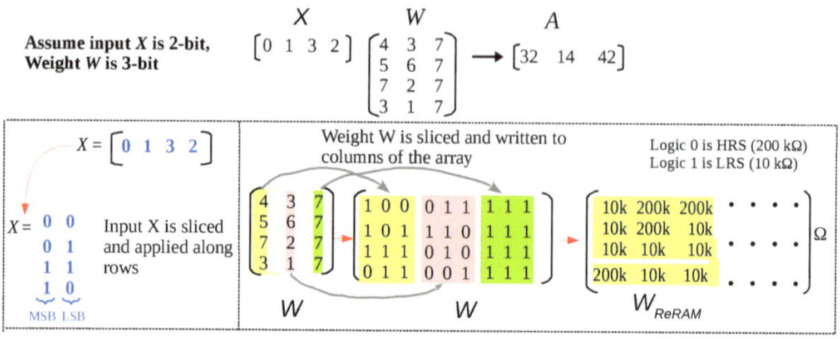

Fig. 8.14 If input vector X is to be multiplied with weight matrix W, each element of X and W are sliced in Bit Slicing approach. Here, each element of X is represented as two bits and each element of W is represented a three bits and written into the array.

4. The next significant bit of all the elements of X are applied to the rows simultaneously.
5. Shift and add the outputs of the column, according to their position and store them in a register (PP_1). Depending on the number of bits in elements of X, Steps 4,5 are repeated. If we have n bits in each element of X, we will have n Partial Products.
6. The partial products corresponding to each bit of X are shifted and added to get the final sum.
   Final Sum $= \sum_{i=0}^{n-1} 2^i \times PP_i$

Let us consider a multiplication of vector X with matrix W where the elements of X are 2-bit integers and W are 3-bit integers, as depicted in Fig.8.14. The steps of MVM using bit slicing for this example is illustrated in Fig.8.15. In this example, both the input X and the weight W are sliced and other variations where only one of them is sliced are possible. Input X is sliced and applied to the rows in two cycles- first the LSB of all its elements followed by the MSB of all its elements. The elements of the weight matrix W are sliced and represented in columns of the array. Therefore, weight 5 of W matrix is represented in three columns as 101. Since the weights are now represented in binary form, they can be easily programmed to two resistive states *e.g.* HRS of 200 kΩ representing '0' and LRS of 10 kΩ representing '1'.

As depicted in Fig.8.15, the LSB of all the elements of X is applied to the rows simultaneously. Note that BL_1 corresponds to the Most Significant Bit (MSB) and BL_3 corresponds to the Least Significant Bit

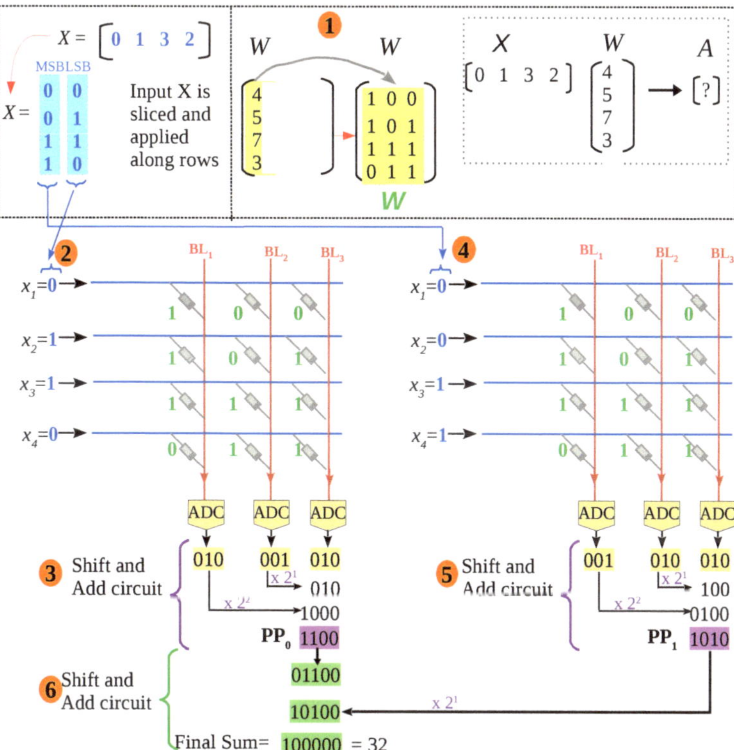

Fig. 8.15 Illustration of MVM $X.W$. Due to space constraints, multiplication of X with only the first column, $W = (4\ 5\ 7\ 3)^T$ is illustrated in 4×3 array here and the remaining columns are not shown.

(LSB) of weight W. Therefore the output of BL_1 needs to be multiplied by 2^2 while the output of BL_3 needs to be multiplied by 2^0. This multiplication to the power of 2 is accomplished by left-shift operation. The shifted results of BL_1, BL_2, BL_3 are then added to get the Partial Product PP_0. A Shift-and-Add circuit is used for each column of the weight matrix to perform this operation. Then, the next significant bit of sliced X is applied to the rows and a similar Shift-and-Add procedure is used to calculate PP_1. Finally, the Partial Products corresponding to each bit of X are weighted according to their position and added. In this example, Final Sum = $PP_0 + 2^1 \times PP_1$ since X is 2-bit integer.

In Fig.8.15, we have used binary representation for MVM for ease of illustration. In reality, the multiplication of each bit of X with each bit of W happens in an analog manner. Fig.8.16 illustrates how the bit value is applied to the rows of the memory array. Since we do not want to disturb the state of the cell while performing MVM, a bit value of '1' is usually translated as a low voltage of 0.2 V or 0.3 V (usually the READ voltage of the ReRAM technology) and a bit value of '0' is translated as just 0 V. Hence (0 1 1 0) at the row is applied as (0 0.2 0.2 0) Volts. If the cell is '0', it is 200 kΩ and we have a current of 1 μA across the cell when the WL is biased at 0.2 V. If the cell is '1', it is 10 kΩ and we have a current of 20 μA across the cell (assuming the BL is held at virtual ground as before). In this manner, multiplication of bit '1' $\times$ '1' results in 20 μA (representing '1') and multiplication of bit '1' $\times$ '0' results in 1 μA (representing '0'). If the size of the array is reasonable (128$\times$128), these currents get added and a column with more '1'$\times$'1' will result in a larger current. The added BL current is fed to the I-to-V converter and converted to a digital output using the ADC. For the 4$\times$3 array depicted in Fig.8.16, a 3-bit ADC is enough since the maximum value of a column is 4 (1+1+1+1). Notice that we need a simpler ADC (3-bit instead of the 6-bit ADC used in Fig.8.13) since each memory cell represents a single bit of weight W. Furthermore, the DAC used in Fig.8.16 is also less complicated since it needs to apply only a single voltage (0.2 V). In this manner, bit slicing reduces the complexity of the peripheral ADCs and DACs.

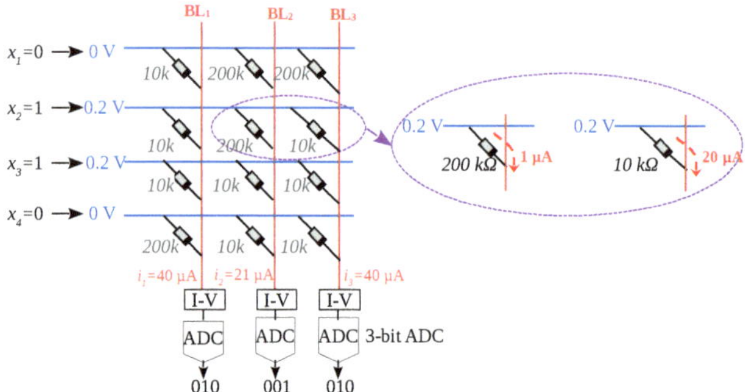

Fig. 8.16 During MVM, in the memory array, the bits are represented as analog values, X as a voltage and W as a resistance.

Since the ReRAM cells are programmed to only two states (HRS and LRS), it can generally be accomplished in a precise manner compared to programming the cells to 2^n states (as needed in pure analog MVM discussed earlier). Therefore the error incurred during MVM due to imprecise programming of the resistive states is overcome. Consequently bit slicing can overcome the accuracy limitation of pure analog NVM to a certain extent. The requirement of binary programming of the states also make the peripheral circuitry (WRITE circuit) a lot less complicated. However, extra shift-and-add circuits are required and they could increase the energy consumption and area of the peripheral circuitry. Note that a shift-and-add circuit is needed for each column of the weight matrix W (only a single column is depicted in Fig.8.15). Finally, latency of MVM using bit slicing is increased when compared to pure analog MVM (Fig.8.13) since the input vector X is sliced to n bits and this requires n multiply operations in the memory array followed by shift-and-add operations.

8.4 Conclusion and Future Outlook

One must be cautioned that all processing should not be moved to memory since it is not always beneficial. There are applications where in-memory computing performs worse when compared to pure-CMOS implementation. For example, performing addition and multiplication in memory in a digital fashion using Boolean logic operations may incur hundreds and few thousands memory cycles as the bit-width increases from 32-bit to 64-bit [22, 62]. In this case, in-memory addition/multiplication makes sense only when the combined cost(energy and delay) to move the inputs from the memory to the processor and compute in a CMOS-based processor is more than the cost of in-memory addition. What is really worth pursuing is a hybrid approach- perform parts of the processing in CMOS (when it is 'better' to compute there) and remaining parts in the memory (when it is 'better' to compute there).

To give an example from the automobile world, computing in future should take an approach similar to Toyota's hybrid technology for driving. With Prius and its successors, Toyota pioneered using both an internal combustion engine and an electric motor. Since the combustion engines have an optimum temperature to which they must be heated (before they can become efficient), the car begins to start moving with the electric motor in a pure electric mode. Then, as the speed increases it switches to combustion engine. If the car is slowed down due to a traffic jam, its switches to electric mode automatically till the queue is cleared. In this manner, the hybrid technology uses the best of both worlds to achieve very good fuel efficiency. The key to achieving fuel efficiency is to exploit the strength of the different technologies at the right time. To move the vehicle, a high torque is needed which an electric motor provides very easily. Using pure electric drive at low speeds is also beneficial for reducing pollution in cities and city limits. On the other hand, range anxiety (number of kms that can be driven before battery loses charge) and the high electric power consumption at high speeds make pure electric drive disadvantageous at high speeds. For high speeds and long distance driving, the combustion engine is advantageous over pure electric drive¶.

The crucial question to be answered is: how do we know when it is 'better' to compute in CMOS technology? and when it is 'better' to compute in the memory array? This question is not easily answered but depends on the application and also on the specific type of NVM technology considered. It must be noted that different NVMs vary widely in their characteristics like endurance, retention, switching time *etc.* Generally, in applications where there is heavy data movement between processor and memory, computing in memory is proven to be energy efficient [6]. If the computing in ReRAM array is performed in an analog fashion like the MVM considered in this chapter, the accuracy is degraded due to cycle-to-cycle and device-to-device variability of ReRAM cells. Therefore, MVM in ReRAM array in an analog manner may not be suitable in applications like scientific computing where a high degree of accuracy is required. On the other hand, there are applications like image processing where an approximate solution is considered acceptable. This is because, a higher accuracy in certain image-processing tasks, even if achievable, is not recognisable to the human eye. In conclusion, computing of future should take a hybrid approach and exploit the advantages of CMOS technology and NVM technology judiciously. The future of computing with the advent of ReRAM and other similar NVM technology will be exciting.

References

[1] A. Pedram, S. Richardson, M. Horowitz, S. Galal, and S. Kvatinsky, "Dark memory and accelerator-rich system optimization in the dark silicon era,"

¶ In reality, Toyota's hybrid technology is more complex since it uses the synergy between electric and combustion technology and there are situations when both work **together** to optimize fuel. However, the purpose of using Toyota's hybrid car analogy here was to drive home the point that computing needs to be performed in a hybrid manner depending on whether it is energy-efficient to be performed in CMOS technology or in the memory array.

IEEE Design Test, vol. 34, pp. 39–50, April 2017.

[2] O. Mutlu, S. Ghose, J. Gómez-Luna, and R. Ausavarungnirun, "Processing data where it makes sense: Enabling in-memory computation," *Microprocessors and Microsystems*, vol. 67, pp. 28–41, 2019.

[3] W. J. Dally, Y. Turakhia, and S. Han, "Domain-specific hardware accelerators," *Commun. ACM*, vol. 63, p. 48–57, jun 2020.

[4] G. Molas and E. Nowak, "Advances in emerging memory technologies: From data storage to artificial intelligence," *Applied Sciences*, vol. 11, no. 23, 2021.

[5] D. Ielmini and H.-S. P. Wong, "In-memory computing with resistive switching devices," *Nature Electronics*, vol. 1, pp. 333 – 343, 2018.

[6] A. Sebastian, M. L. Gallo, R. Khaddam-Aljameh, and E. Eleftheriou, "Memory devices and applications for in-memory computing," *Nature Nanotechnology*, 2020.

[7] H. A. D. Nguyen, J. Yu, M. A. Lebdeh, M. Taouil, S. Hamdioui, and F. Catthoor, "A classification of memory-centric computing," *J. Emerg. Technol. Comput. Syst.*, vol. 16, Jan. 2020.

[8] N. Xu, T. Park, K. J. Yoon, and C. S. Hwang, "In-memory stateful logic computing using memristors: Gate, calculation, and application," *physica status solidi (RRL) – Rapid Research Letters*, vol. 15, no. 9, p. 2100208, 2021.

[9] Y. S. Kim, M. W. Son, and K. M. Kim, "Memristive stateful logic for edge boolean computers," *Advanced Intelligent Systems*, vol. 3, no. 7, p. 2000278, 2021.

[10] S. Kvatinsky, G. Satat, N. Wald, E. G. Friedman, A. Kolodny, and U. C. Weiser, "Memristor-based material implication (imply) logic: Design principles and methodologies," *IEEE Transactions on Very Large Scale Integration (VLSI) Systems*, vol. 22, pp. 2054–2066, Oct 2014.

[11] L. Amarú, P. E. Gaillardon, and G. D. Micheli, "Majority-inverter graph: A new paradigm for logic optimization," *IEEE Transactions on Computer-Aided Design of Integrated Circuits and Systems*, vol. 35, pp. 806–819, May 2016.

[12] J. Reuben, R. Ben-Hur, N. Wald, N. Talati, A. Ali, P.-E. Gaillardon, and S. Kvatinsky, "Memristive logic: A framework for evaluation and comparison," in *Power And Timing Modeling, Optimization and Simulation (PATMOS)*, pp. 1–8, September 2017.

[13] E. Lehtonen, J. H. Poikonen, and M. Laiho, "Memristive stateful logic," in *Memristor Networks* (A. Adamatzky and L. Chua, eds.), pp. 603–623, Cham: Springer International Publishing, 2014.

[14] S. Shirinzadeh, M. Soeken, P. Gaillardon, and R. Drechsler, "Fast logic synthesis for rram-based in-memory computing using majority-inverter graphs," in *2016 Design, Automation Test in Europe Conference Exhibition (DATE)*, pp. 948–953, 2016.

[15] S. Li, C. Xu, Q. Zou, J. Zhao, Y. Lu, and Y. Xie, "Pinatubo: A processing-in-memory architecture for bulk bitwise operations in emerging non-volatile memories," in *Proceedings of the 53rd Annual Design Automation Conference*, DAC '16, (New York, NY, USA), pp. 173:1–173:6, ACM, 2016.

[16] S. Kvatinsky, D. Belousov, S. Liman, G. Satat, N. Wald, E. G. Friedman, A. Kolodny, and U. C. Weiser, "Magic—memristor-aided logic," *IEEE Transactions on Circuits and Systems II: Express Briefs*, vol. 61, no. 11, pp. 895–899, 2014.

[17] B. Hoffer, V. Rana, S. Menzel, R. Waser, and S. Kvatinsky, "Experimental demonstration of memristor-aided logic (magic) using valence change memory (vcm)," *IEEE Transactions on Electron Devices*, vol. 67, no. 8, pp. 3115–3122, 2020.

[18] E. Testa, M. Soeken, L. G. Amaru, and G. De Micheli, "Logic synthesis for established and emerging computing," *Proceedings of the IEEE*, vol. 107, no. 1, pp. 165–184, 2019.

[19] G. Jaberipur, B. Parhami, and D. Abedi, "Adapting computer arithmetic structures to sustainable supercomputing in low-power, majority-logic nanotechnologies," *IEEE Transactions on Sustainable Computing*, vol. 3, pp. 262–273, Oct 2018.

[20] V. Pudi, K. Sridharan, and F. Lombardi, "Majority logic formulations for parallel adder designs at reduced delay and circuit complexity," *IEEE Transactions on Computers*, vol. 66, pp. 1824–1830, Oct 2017.

[21] J. Reuben and S. Pechmann, "A parallel-friendly majority gate to accelerate in-memory computation," in *2020 IEEE 31st International Conference on Application-specific Systems, Architectures and Processors (ASAP)*, pp. 93–100, 2020.

[22] J. Reuben and S. Pechmann, "Accelerated addition in resistive ram array using parallel-friendly majority gates," *IEEE Transactions on Very Large Scale Integration (VLSI) Systems*, vol. 29, no. 6, pp. 1108–1121, 2021.

[23] J. Reuben, "Binary addition in resistance switching memory array by sensing majority," *Micromachines*, vol. 11, no. 5, 2020.

[24] S. Pechmann, T. Mai, M. Völkel, M. K. Mahadevaiah, E. Perez, E. Perez-Bosch Quesada, M. Reichenbach, C. Wenger, and A. Hagelauer, "A versatile, voltage-pulse based read and programming circuit for multi-level rram cells," *Electronics*, vol. 10, no. 5, 2021.

[25] H. Padberg, A. Regev, G. Piccolboni, A. Bricalli, G. Molas, J. F. Nodin, and S. Kvatinsky, "Experimental demonstration of non-stateful in-memory logic with 1t1r oxram valence change mechanism memristors," *IEEE Transactions on Circuits and Systems II: Express Briefs*, pp. 1–1, 2023.

[26] L. Brackmann, T. Ziegler, A. Jafari, D. J. Wouters, M. B. Tahoori, and S. Menzel, "Improved arithmetic performance by combining stateful and non-stateful logic in resistive random access memory 1t–1r crossbars," *Advanced Intelligent Systems*, vol. 6, no. 3, p. 2300579, 2024.

[27] R. Ben Hur, N. Wald, N. Talati, and S. Kvatinsky, "Simple magic: Synthesis and in-memory mapping of logic execution for memristor-aided logic," in *2017 IEEE/ACM International Conference on Computer-Aided Design (ICCAD)*, pp. 225–232, 2017.

[28] R. Ben-Hur, R. Ronen, A. Haj-Ali, D. Bhattacharjee, A. Eliahu, N. Peled, and S. Kvatinsky, "Simpler magic: Synthesis and mapping of in-memory logic executed in a single row to improve throughput," *IEEE Transactions on Computer-Aided Design of Integrated Circuits and Systems*, vol. 39, no. 10, pp. 2434–2447, 2020.

[29] Rumi Zhang, K. Walus, Wei Wang, and G. A. Jullien, "A method of majority logic reduction for quantum cellular automata," *IEEE Transactions on Nanotechnology*, vol. 3, pp. 443–450, Dec 2004.

[30] S. Sheu, K. Cheng, M. Chang, P. Chiang, W. Lin, H. Lee, P. Chen, Y. Chen, T. Wu, F. T. Chen, K. Su, M. Kao, and M. Tsai, "Fast-write resistive ram (rram) for embedded applications," *IEEE Design Test of Computers*, vol. 28, pp. 64–71, Jan 2011.

[31] P. Huang, J. Kang, Y. Zhao, S. Chen, R. Han, Z. Zhou, Z. Chen, W. Ma, M. Li, L. Liu, and X. Liu, "Reconfigurable nonvolatile logic operations in resistance switching crossbar array for large-scale circuits," *Advanced Materials*, vol. 28, no. 44, pp. 9758–9764, 2016.

[32] A. Karimi and A. Rezai, "Novel design for a memristor-based full adder using a new imply logic approach," *J Comput Electron*, vol. 17, p. 11303–1314, 2018.

[33] L. Cheng *et al.*, "Functional demonstration of a memristive arithmetic logic unit (memalu) for in-memory computing," *Advanced Functional Materials*, vol. 29, no. 49, p. 1905660, 2019.

[34] S. Ganjeheizadeh Rohani, N. Taherinejad, and D. Radakovits, "A semiparallel full-adder in imply logic," *IEEE Transactions on Very Large Scale Integration (VLSI) Systems*, vol. 28, no. 1, pp. 297–301, 2020.

[35] N. Talati, S. Gupta, P. Mane, and S. Kvatinsky, "Logic design within memristive memories using memristor-aided logic (magic)," *IEEE Transactions on Nanotechnology*, vol. 15, no. 4, pp. 635–650, 2016.

[36] Y. S. Kim, M. W. Son, H. Song, J. Park, J. An, J. B. Jeon, G. Y. Kim, S. Son, and K. M. Kim, "Stateful in-memory logic system and its practical implementation in a taox-based bipolar-type memristive crossbar array," *Advanced Intelligent Systems*, vol. 2, no. 3, p. 1900156, 2020.

[37] K. Alhaj Ali *et al.*, "Memristive computational memory using memristor overwrite logic (mol)," *IEEE Transactions on Very Large Scale Integration (VLSI) Systems*, vol. 28, no. 11, pp. 2370–2382, 2020.

[38] A. Siemon, R. Drabinski, M. J. Schultis, X. Hu, E. Linn, A. Heittmann, R. Waser, D. Querlioz, S. Menzel, and J. S. Friedman, "Stateful three-input logic with memristive switches," *Scientific Reports*, vol. 9, no. 1, p. 14618, 2019.

[39] A. Siemon, S. Menzel, R. Waser, and E. Linn, "A complementary resistive switch-based crossbar array adder," *IEEE Journal on Emerging and Selected Topics in Circuits and Systems*, vol. 5, no. 1, pp. 64–74, 2015.

[40] N. TaheriNejad, "Sixor: Single-cycle in-memristor xor," *IEEE Transactions on Very Large Scale Integration (VLSI) Systems*, pp. 1–11, 2021.

[41] F. Pinto and I. Vourkas, "Robust circuit and system design for general-purpose computational resistive memories," *Electronics*, vol. 10, no. 9, 2021.

[42] Z. Wang *et al.*, "Efficient implementation of boolean and full-adder functions with 1t1r rrams for beyond von neumann in-memory computing," *IEEE Transactions on Electron Devices*, vol. 65, pp. 4659–4666, Oct 2018.

[43] A. Siemon *et al.*, "Sklansky tree adder realization in 1s1r resistive switching memory architecture," *The European Physical Journal Special Topics*, vol. 228, no. 10, pp. 2269–2285, 2019.

[44] J. Reuben, "Rediscovering majority logic in the post-cmos era: A perspective from in-memory computing," *Journal of Low Power Electronics and Applications*, vol. 10, no. 3, 2020.

[45] P. Wang, M. Y. Niamat, S. R. Vemuru, M. Alam, and T. Killian, "Synthesis of majority/minority logic networks," *IEEE Transactions on Nanotechnology*, vol. 14, no. 3, pp. 473–483, 2015.

[46] C.-C. Chung, Y.-C. Chen, C.-Y. Wang, and C.-C. Wu, "Majority logic circuits optimisation by node merging," in *2017 22nd Asia and South Pacific Design Automation Conference (ASP-DAC)*, pp. 714–719, 2017.

[47] H. Riener, E. Testa, L. Amaru, M. Soeken, and G. D. Micheli, "Size optimization of migs with an application to qca and stmg technologies," in *2018 IEEE/ACM International Symposium on Nanoscale Architectures (NANOARCH)*, pp. 1–6, 2018.

[48] R. Devadoss, K. Paul, and M. Balakrishnan, "Majority logic: Prime implicants and n-input majority term equivalence," in *2019 32nd International Conference on VLSI Design and 2019 18th International Conference on Embedded Systems (VLSID)*, pp. 464–469, 2019.

[49] A. Neutzling, F. S. Marranghello, J. M. Matos, A. Reis, and R. P. Ribas, "maj-n logic synthesis for emerging technology," *IEEE Transactions on Computer-Aided Design of Integrated Circuits and Systems*, vol. 39, no. 3, pp. 747–751, 2020.

[50] M. Kaneko, "A novel framework for procedural construction of parallel prefix adders," in *2019 IEEE International Symposium on Circuits and Systems (ISCAS)*, pp. 1–5, 2019.

[51] C. L. Ayala, N. Takeuchi, Y. Yamanashi, T. Ortlepp, and N. Yoshikawa, "Majority-logic-optimized parallel prefix carry look-ahead adder families using adiabatic quantum-flux-parametron logic," *IEEE Transactions on Applied Superconductivity*, vol. 27, pp. 1–7, June 2017.

[52] J. Reuben, "Design of in-memory parallel-prefix adders," *Journal of Low Power Electronics and Applications*, vol. 11, no. 4, 2021.

[53] A. Levisse, B. Giraud, J. . Noel, M. Moreau, and J. . Portal, "Rram crossbar arrays for storage class memory applications: Throughput and density considerations," in *2018 Conference on Design of Circuits and Integrated Systems (DCIS)*, pp. 1–6, Nov 2018.

[54] A. Amirsoleimani, F. Alibart, V. Yon, J. Xu, M. R. Pazhouhandeh, S. Ecoffey, Y. Beilliard, R. Genov, and D. Drouin, "In-memory vector-matrix multiplication in monolithic complementary metal–oxide–semiconductor-memristor integrated circuits: Design choices, challenges, and perspectives," *Advanced Intelligent Systems*, vol. 2, no. 11, p. 2000115, 2020.

[55] R. Nägele, J. Finkbeiner, V. Stadtlander, M. Grözing, and M. Berroth, "Analog multiply-accumulate cell with multi-bit resolution for all-analog ai inference accelerators," *IEEE Transactions on Circuits and Systems I: Regular Papers*, vol. 70, no. 9, pp. 3509–3521, 2023.

[56] Z. Sun and D. Ielmini, "Invited tutorial: Analog matrix computing with crosspoint resistive memory arrays," *IEEE Transactions on Circuits and Systems II: Express Briefs*, vol. 69, no. 7, pp. 3024–3029, 2022.

[57] M. Hu, C. E. Graves, C. Li, Y. Li, N. Ge, E. Montgomery, N. Davila, H. Jiang, R. S. Williams, J. J. Yang, Q. Xia, and J. P. Strachan, "Memristor-based analog computation and neural network classification with a dot product engine," *Advanced Materials*, vol. 30, no. 9, p. 1705914, 2018.

[58] S. Agarwal, T.-T. Quach, O. Parekh, A. H. Hsia, E. P. DeBenedictis, C. D. James, M. J. Marinella, and J. B. Aimone, "Energy scaling advantages of resistive memory crossbar based computation and its application to sparse coding," *Frontiers in Neuroscience*, vol. 9, 2016.

[59] K. Roy, I. Chakraborty, M. Ali, A. Ankit, and A. Agrawal, "In-memory computing in emerging memory technologies for machine learning: An overview," in *Proceedings of the 57th ACM/EDAC/IEEE Design Automation Conference*, DAC '20, IEEE Press, 2020.

[60] Y. Luo, S. Wang, P. Zuo, Z. Sun, and R. Huang, "Modeling and mitigating the interconnect resistance issue in analog rram matrix computing circuits," *IEEE Transactions on Circuits and Systems I: Regular Papers*, vol. 69, no. 11, pp. 4367–4380, 2022.

[61] A. Shafiee, A. Nag, N. Muralimanohar, R. Balasubramonian, J. P. Strachan, M. Hu, R. S. Williams, and V. Srikumar, "Isaac: a convolutional neural network accelerator with in-situ analog arithmetic in crossbars," *SIGARCH Comput. Archit. News*, vol. 44, p. 14–26, June 2016.

[62] V. Lakshmi, J. Reuben, and V. Pudi, "A novel in-memory wallace tree multiplier architecture using majority logic," *IEEE Transactions on Circuits and Systems I: Regular Papers*, vol. 69, no. 3, pp. 1148–1158, 2022.

Also by John Reuben

Non-Volatile Memories: Device and Peripheral Circuit Design

Covering Flash, Resistive-switching, and Ferroelectric Memories

(Expected Publication: June 2026)

This comprehensive book is written for beginners to introduce them to non-volatile memories, which will form an integral part of future computing systems. AI hardware will be predominantly memory-centric with computation being performed in memory or near memory arrays. After briefly introducing SRAM and DRAM, this book will introduce the reader to floating gate MOSFET memory (flash), the current standard for non-volatile memories. Then, this book will discuss Resistive-switching memories (ReRAM, Phase Change Memory, STT/SOT-MRAM) and ferroelectric memories (FeRAM, FTJ, FeFET). For each of these memories, the device stucture, array configuration and READ and WRITE process are described in detail with suitable circuits. The book concludes with a chapter on in-memory computing so that readers can appreciate how these memories are used for Vector Matrix Multiplication, the fundamental computation in Deep Neural Networks.

www.ingramcontent.com/pod-product-compliance
Ingram Content Group UK Ltd.
Pitfield, Milton Keynes, MK11 3LW, UK
UKHW060359300726
14090UKWH00001B/28